Lecture Notes in Computer Scie

Commenced Publication in 1973
Founding and Former Series Editors:
Gerhard Goos, Juris Hartmanis, and Jan van Leeuwen

Lei Chen   Chengfei Liu   Xiao Zhang
Shan Wang   Darijus Strasunskas
Stein L. Tomassen   Jinghai Rao
Wen-Syan Li   K. Selçuk Candan
Dickson K.W. Chiu   Yi Zhuang
Clarence A. Ellis   Kwang-Hoon Kim (Eds.)

Advances in Web and Network Technologies, and Information Management

APWeb/WAIM 2009 International Workshops:
WCMT 2009, RTBI 2009, DBIR-ENQOIR 2009, PAIS 2009
Suzhou, China, April 2-4, 2009
Revised Selected Papers

 Springer

Volume Editors

Lei Chen
Hong Kong University of Science and Technology, E-mail: leichen@cse.ust.hk
Chengfei Liu
Swinburne University of Technology, E-mail: cliu@groupwise.swin.edu.au
Xiao Zhang
Renmin Universty of China, E-mail: zhangxiao@ruc.edu.cn
Shan Wang
Renmin University China, E-mail: swang@ruc.edu.cn
Darijus Strasunskas
NTNU, Norway, E-mail: darijuss@gmail.com
Stein L. Tomassen
NTNU Norway, E-mail: stein.l.tomassen@idi.ntnu.no
Jinghai Rao
AOL, China, E-mail: raojinghai@gmail.com
Wen-Syan Li
SAP Research China, E-mail: wen-syan.li@sap.com
K. Selçuk Candan
Arizona State University, E-mail: candan@asu.edu
Dickson K.W. Chiu
Dickson Computer Systems,Hong Kong, E-mail: dicksonchiu@ieee.org
Yi Zhuang
Zhejiang Gongshang University, China, E-mail: zhuang@zjgsu.edu.cn
Clarence A. Ellis
University of Colorado at Boulder, E-mail: sip@cs.colorado.edu
Kwang-Hoon Kim
Kyonggi University, Korea, E-mail: kwang@kyonggi.ac.kr

Library of Congress Control Number: 2009932910

CR Subject Classification (1998): H.2.8, H.5, H.3.5, H.2, K.8.1, D.4.2

LNCS Sublibrary: SL 3 – Information Systems and Application, incl. Internet/Web
and HCI

ISSN   0302-9743
ISBN-10   3-642-03995-2 Springer Berlin Heidelberg New York
ISBN-13   978-3-642-03995-9 Springer Berlin Heidelberg New York

springer.com

© Springer-Verlag Berlin Heidelberg 2009
Printed in Germany

Typesetting: Camera-ready by author, data conversion by Scientific Publishing Services, Chennai, India
Printed on acid-free paper   SPIN: 12736430   06/3180   5 4 3 2 1 0

APWeb/WAIM 2009 Workshop Chair's Message

APWeb and WAIM are leading international conferences on research, development and applications of Web technologies, database systems, information management and software engineering, with a focus on the Asia-Pacific region. Previous APWeb conferences were held in Beijing (1998), Hong Kong (1999), Xian (2000), Changsha (2001), Xian (2003), Hangzhou (2004), Shanghai (2005), Harbin (2006), Huangshan (2007) and Shenyang (2008) and previous WAIM conferences were held in Shanghai (2000), Xian (2001), Beijing (2002), Chengdu (2003), Dalian (2004), Hangzhou (2005), Hong Kong (2006), Huangshan (2007) and Zhangjiajie (2008).

For the second time, APWeb and WAIM were combined to foster closer collaboration and research idea sharing. In 2007, APWeb and WAIM were first combined and had co-located excellent workshops on specialized and emerging topics. To continue this excellent program, we were pleased to serve as the Workshop Chairs for APWeb-WAIM 2009.

This volume comprises papers from four APWeb-WAIM 2009 workshops:

1. International Workshop on Web-Based Contents Management Technologies (WCMT 2009)
2. International Workshop on Real-Time Business Intelligence (RTBI 2009)
3. International Workshop on DataBase and Information Retrieval and Aspects in Evaluating Holistic Quality of Ontology-Based Information Retrieval (DBIR-ENQOIR 2009)
4. International Workshop on Process-Aware Information Systems (PAIS 2009)

These four workshops were selected from a public call-for-proposals process. The workshop organizers put a tremendous amount of effort into soliciting and selecting research papers with a balance of high quality and new ideas and new applications. We asked all workshops to follow a rigid paper selection process, including the procedure to ensure that any Program Committee members (including Workshop Program Committee Chairs) are excluded from the paper review process of any papers they are involved in. A requirement about the overall paper acceptance rate was also imposed on all the workshops.

We are very grateful to Qing Li, Ling Feng, Jian Pei, and Jianxi Fan and many other people for their great effort in supporting the conference organization. We would like to take this opportunity to thank all workshop organizers and Program Committee members for their great effort to put together the workshop program of APWeb-WAIM 2009. Last but not least, we thank the generous funding support from the Future Network Centre (FNC) of the City University of Hong Kong under CityU Applied R&D Centre (Shenzhen), through grant number 9681001.

April 2009

Lei Chen
Chengfei Liu

International Workshop on Real-Time Business Intelligence (RTBI 2009) Chairs' Message

The first workshop on Real-Time Business Intelligence (RTBI 09), co-located with the APWEB/WAIM 2009 conference, focused on the challenges associated with the modeling, architectural designs, technologies, and applications for real-time business intelligence. As one of the most critical tools in obtaining competitive advantage (and thus surviving) in the business world, real-time business intelligence is one of the key driving forces in IT investments. Reliable and real-time collection of business intelligence requires co-operation among the many loosely coupled components of the IT infrastructure of an enterprise. Consequently, as these infrastructures become more complex and as the data sizes reach the petabyte levels, designing and deploying real-time business intelligence solutions is becoming a real challenge. In this workshop, we explored alternative solutions to the imminent "scalability" problem in real-time business intelligence.

April 2009

Wen-Syan Li
K. Selcuk Candan

International Workshop on Web-Based Contents Management Technologies (WCMT 2009) Chairs' Message

The Web is ever-increasing in size and involving a quite broad range of technologies such as databases, XML and digital contents, data and Web mining, Web services, Semantic Web and ontology, information retrieval, and others. WCMT 2009 was held in conjunction with the WAIM/APWeb 2009 conference and invited original research contributions on XML, Web, and Internet contents technologies. WCMT 2009 aimed at bringing together researchers in different fields related to Web information processing who have a common interest in interdisciplinary research. The workshop provided a forum where researchers and practitioners could share and exchange their knowledge and experience.

April 2009

Dickson K.W. Chiu
Yi Zhuang

International Workshop on Process-Aware Information Systems (PAIS 2009) Chairs' Message

Recently, the concept of process has become one of the hottest issues in the Web-age enterprise information technology arena; consequently process-aware information systems (PAIS) have been rapidly spreading their applications over various industry domains. These phenomena come from the strong belief that process-aware information systems help large organizations improve their business operations and management; moreover, these systems have proved to be very effective. Nevertheless, many of the PAIS products have been found to be ineffective in some serious applications due to the lack of functional capability, architectural extensibility and agility for the new requirements coping with many advanced features and functionalities. Therefore, through this workshop, we explored the process-aware information systems domain and looked for feasible solutions for those PAIS products suffering from serious functional and architectural problems.

April 2009

Kwang-Hoon Kim
Clarence A. Ellis

International Workshop on DataBase and Information Retrieval and on Aspects in Evaluating Holistic Quality of Ontology-Based Information Retrieval (DBIR-ENQOIR 2009) Chairs' Message

Information retrieval (IR) is a traditional research field different from the field of DataBase(DB). They both have evolved independently and fruitfully over many decades. However, modern applications, in particular Web-enabled applications, such as customer support, e-business, digital libraries require of search the both structured, or tabular, data and semi-structured or unstructured data. DB and IR on their own lack the functionality to handle the other side. Thus DBIR is emerging as a hot research area. One of popular topics in this area is keyword search over relational databases.

The ENQOIR aims to deeper understanding and to disseminate knowledge on advances in the evaluation and application of ontology-based information retrieval (ObIR). The main areas of the workshop are an overlap between three evaluation aspects in ObIR, namely, evaluation of information retrieval, evaluation of ontology quality impact on ObIR results, and evaluation of user interaction complexity. The main objective is to contribute to the optimization of ObIR by systemizing the existing body of knowledge on ObIR and defining a set of metrics for the evaluation of ontology-based search. The long-term goal of the workshop is to establish a forum to analyze and proceed toward a holistic method for the evaluation of ontology-based information retrieval systems.

April 2009

Xiao Zhang
Shan Wang
Darijus Strasunskas
Stein L. Tomassen
Jinghai Rao

Organization

International Workshop on Real-Time Business Intelligence (RTBI 2009)

Program Co-chairs

Wen-Syan Li   SAP Research China
K. Selcuk Candan   Arizona State University, USA

Program Committee

Howard Ho   IBM Almaden Research Center, USA
Thomas Phan   Microsoft, USA
Aoying Zhou   East China Normal University, China
Walid G. Aref   Purdue University, USA
Divyakant Agrawal UC Santa Barbara, USA
Xuemin Lin   University of New South Wales, Australia
Jun Tatemura   NEC Labs, USA
Daniel Zilio   IBM Toronto Labs, Canada
Huiping Cao   Arizona State University, USA
Nesime Tatbul   ETH Zurich, Switzerland
Fabio Casati   University of Trento, Italy
Malu G.   Castellanos, HP Labs, USA
Xiaohui (Helen) Gu North Carolina State University, USA

International Workshop on Web-Based Contents Management Technologies(WCMT 2009)

General Co-chairs

Jiajin Le   Donghua University, China
Hua Hu   Zhejiang Gongshang University, China

Program Co-chairs

Dickson K.W. Chiu   Dickson Computer System, Hong Kong, China
Yi Zhuang   Zhejiang Gongshang University, China

Program Committee

Jun Yang   Carnegie Mellon University, USA
Weining Qian   East China Normal University, China
Panos Kalnis   National University of Singapore, Singapore
Cuiping Li   Renmin University, China
Yi Chen   Arizona State University, USA
Aibo Song   Southeast University, China
Weili Wu   University of Texas at Dallas
Wei Wang   University of New South Wales, Australia
Fei Wu   Zhejiang University, China
Mustafa Atay   Winston-Salem State University, USA
Jianliang Xu   Hong Kong Baptist University, China
Jie Shao   University of Melbourne, Australia
Raymond Chi-Wing
  Wong   Hong University of Science and Technology,
                             China
Haiyang Hu   Zhejiang Gongshang University, China
Yunjun Gao   Singapore Management University, Singapore
Guojie Song   Beijing University, China
Zi Huang   The Open University, UK
Bingsheng He   Microsoft Research Asia, China
Gao Cong   Aalborg University, Denmark
Sharad Mehrotra   University of California, USA
Feifei Li   Florida State University, USA
Hong Shen   University of Adelaide, Australia
Byron Choi   Hong Kong Baptist University, China

International Workshop on Process-Aware Information Systems (PAIS 2009)

Program Co-chairs

Clarence A. Ellis   University of Colorado at Boulder, USA
Kwang-Hoon Kim   Kyonggi University, Korea

Program Committee

Hayami Haruo   Kanagawa Institute of Technology, Japan
Jintae Lee   University of Colorado at Boulder, USA
George Wyner   Boston University, USA
Jorge Cardoso   University of Madeira, Portugal
Yang Chi-Tsai   Flowring Technology, Inc., Taiwan
Michael zur Muehlen   Stevenson Institute of Technology, USA
Dongsoo Han   Information and Commnications University,
                            Korea
Ilkyeun Ra   University of Colorado at Denver, USA
Taekyou Park   Hanseo University, Korea
Joonsoo Bae   Chonbuk National University, Korea
Yoshihisa Sadakane   NEC Soft, Japan
Junchul Chun   Kyonggi University, Korea
Luis Joyanes Aguilar   Universidad Pontificia de Salamanca, Spain
Tobias Rieke   University of Muenster, Germany
Modrak Vladimir   Technical University of Kosice, Slovakia
Haksung Kim   Dongnam Health College, Korea
Yongjoon Lee   Electronics and Telecommunications Research
                            Institute, Korea
Taekyu Kang   Electronics and Telecommunications Research
                            Institute, Korea
Jinjun Chen   Swinburne University of Technology, Australia
Zongwei Luo   The University of Hong Kong, China
Peter sakal   Technical University of Bratislava, Slovakia
Sarka Stojarova   Univeristy in Brno, Czech Republic
Boo-Hyun Lee   KongJu National University, Korea
Jeong-Hyun Park   Electronics and Telecommunications Research
                            Institute, Korea
Yanbo Han   Chinese Academy of Sciences, China
Jacques Wainer   State University of Campinas, Brazil
Aubrey J. Rembert   University of Colorado at Boulder, USA

International Workshop on DataBase and Information Retrieval and on Aspects in Evaluating Holistic Quality of Ontology-Based Information Retrieval (DBIR-ENQOIR 2009)

Program Co-chairs

Xiao Zhang	Renmin University, China
Shan Wang	Professor, Renmin University, China
Darijus Strasunskas	NTNU, Norway
Stein L. Tomassen	NTNU, Norway
Jinghai Rao	AOL, China

Program Committee

Xuemin Lin	University of New South Wales, Australia
Ken Pu	University of Ontario Institute of Technolgy, Canada
Goce Trajcevski	Northwestern University, USA
Jianyong Wang	Tsing Hua University, China
Ge Yu	Northeastern University, China
Jeffery X. Yu	Chinese University of Hong Kong, China
Xiaohui Yu	York University, Canada
Jiang Zhan	Remin University, China
Jun Zhang	Dalian Maritime University, China
Aoying Zhou	Fudan University, China
Per Gunnar Auran	Yahoo! Technologies, Norway
Xi Bai	University of Edinburgh, UK
Robert Engels	ESIS, Norway
Avigdor Gal	Technion, Israel
Jon Atle Gulla	NTNU, Norway
Sari E. Hakkarainen	Finland
Monika Lanzenberger	Vienna University of Technology, Austria
Kin Fun Li	University of Victoria, Canada
Federica Mandreoli	University of Modena e Reggio Emilia, Italy
James C. Mayfield	John Hopkins University, USA
Gabor Nagypal	disy Informationssysteme GmbH, Germany
David Norheim	Computas, Norway
Jaana Kekalainen	University of Tampere, Finland
Iadh Ounis	University of Glasgow, UK
Marta Sabou	The Open University, UK
Tetsuya Sakai	NewsWatch, Inc., Japan
Amanda Spink	Queensland University of Technology, Australia
Peter Spyns	Vrije Universiteit Brussel, Belgium
Heiko Stoermer	University of Trento, Italy
Victoria Uren	The Open University, UK

Table of Contents

International Workshop on Process Aware Information Systems (PAIS 2007)

PAIS

International Workshop on Data Base and Information Retrieval and on Aspects in Evaluating Holistic Quality of Ontology-Based Information Retrieval (DBIR-ENQOIR 2009)

DBIR-ENQOIR 1

DBIR-ENQOIR 2

ScaMMDB: Facing Challenge of Mass Data Processing with MMDB[*]

Yansong Zhang[1,2,3], Yanqin Xiao[1,2,4], Zhanwei Wang[1,2], Xiaodong Ji[1,2], Yunkui Huang[1,2], and Shan Wang[1,2]

[1] Key Laboratory of the Ministry of Education for Data Engineering and Knowledge Engineering, Renmin University of China, Beijing 100872, China
[2] School of Information, Renmin University of China, Beijing 100872, China
[3] Department of Computer Science, Harbin Financial College, Harbin 150030, China
[4] Computer Center, Hebei University, Baoding Hebei 071002, China
{zhangys_ruc,xyqwang,swang }@ruc.edu.cn

Abstract. Main memory database(MMDB) has much higher performance than disk resident database(DRDB), but the architecture of hardware limits the scalability of memory capacity. In OLAP applications, comparing with data volume, main memory capacity is not big enough and it is hard to extend. In this paper, ScaMMDB prototype is proposed towards the scalability of MMDB. A multi-node structure is established to enable system to adjust total main memory capacity dynamically when new nodes enter the system or some nodes leave the system. ScaMMDB is based on open source MonetDB which is a typical column storage model MMDB, column data transmission module, column data distribution module and query execution plan re-writing module are developed directly in MonetDB. Any node in ScaMMDB can response user's requirements and SQL statements are transformed automatically into extended column operating commands including local commands and remote call commands. Operation upon certain column is pushed into the node where column is stored, current node acts as temporarily mediator to call remote commands and assembles the results of each column operations. ScaMMDB is a test bed for scalability of MMDB, it can extend to MMDB cluster, MMDB replication server, even peer-to-peer OLAP server for further applications.

1 Introduction

Data accessing in main memory is much faster than data accessing on disk. More and more MMDBs enter enterprise applications such as TimesTen, solidDB and Altibase et al., most MMDB systems are employed in high performance real-time processing

[*] Supported by the National Natural Science Foundation of China under Grant No. 60473069,60496325; the joint research of HP Lab China and Information School of Renmin University(Large Scale Data Management);the joint research of Beijing Municipal Commission of education and Information School of Renmin University(Main Memory OLAP Server); the Renmin University of China Graduate Science Foundation No. 08XNG040.

L. Chen et al. (Eds.): APWeb and WAIM 2009, LNCS 5731, pp. 1–12, 2009.
© Springer-Verlag Berlin Heidelberg 2009

areas for example telecom applications. The advanced hardware technique supports large capacity memory(from GB to TB memory) for high-end servers, but comparing with the data volume of large database for example the top ten databases[1] in the world, the capacity of memory is far below the data volume. The capacity of memory can not increase as data volume increases because of the interface limitation on motherboard or hardware compatibility problem.

There are some limitations with scalability of MMDB servers:

1.   High-end servers with large memory capacity support are much expensive;
2.   When updating servers, the memory capacity is limited by the amount of RAM slots, the model of memory bank, the parameters of memory bank(e.g. memory frequency, working voltage etc);
3.   Non-stop applications demand more smoothly memory updating mode while adding memory bank on motherboard has to power off the server;
4.   Memory is the new performance bottleneck in MMDB system, cache capacity can not enlarge as memory capacity increases, so the performance gap between cache and memory becomes larger. The scalability not only focuses on memory capacity but also focuses on processing capacity.

Our research work focuses on MMDB in OLAP areas which involve a large amount of complicated queries with large computing workloads. We propose a loosely managed MMDB server network named as ScaMMDB which takes the advantage of high speed network to form a virtual "network memory" for enlarging the total memory capacity of the system. Fiber optic network operates at high speed from Gb/s to Tb/s, it is much faster than the speed of data accessing from hard disk limited by the rotating speed of disks. Each server node shares its memory just as a secondary main memory of query responding node. Furthermore, each node has data processing capability; they can either act as dumb memory sharing node or act as data processing node to process the "pushed down" operations.

Our contributions are as follows:

1.   We propose a multi-node based DSM MMDB system, which can dynamically enlarge the total memory capacity to adapt to the data volume requirement of large database.
2.   We use high speed network to extend traditional memory, MMDB nodes share their memories through network to form a virtual "netMemory" which can provide higher performance over hard disk and acts as the auxiliary main memory.
3.   We establish a prototype based on open source MMDB MonetDB which is a high performance MMDB for OLAP applications. We extend the low level module and APIs to enable data sharing among multiple nodes and executing queries just like a single MonetDB server.

The background notions and conventions are presented in Sec.2. Architecture of ScaMMDB is discussed in Sec. 3. Sec.4 shows the core modules of system. In Sec.5 we show our experiment results. Finally, Sec.6 summarizes the paper and discusses future work.

2 Preliminaries

We design experiments to test the effect of performance with different memory capacity conditions. We select MonetDB as our testing target which has higher performance in TPC-H testing than DRDBs and other typical row storage MMDB. We gradually increase data volume at granularity of 25% memory capacity, the power@size measure of TPC-H shows that it drops as data volume increases. When data volume exceeds 75% of memory capacity, power@size value rapidly decreases till system crash when data volume exceeds total capacity of memory and swap. So the high performance of MMDB relies on sufficient memory supply. When data volume

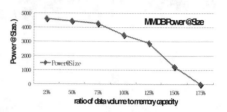

Fig. 1. TPC-H Power@Size

exceeds maximum memory capacity that server can support, how can we obtain additional memory resource without updating hardware?

Our research focuses on MMDB OLAP applications which involve a lot of high selectivity clauses, joins and aggregate operations, the computations are usually costly but the query results are always small. Column storage(DSM) is superior to row-wise storage(NSM) in OLAP scenarios[5-7], our experiment results also show that MonetDB which is DSM based has higher TPC-H measure values than other candidate systems. Naturedly, we consider how to extend memory capacity by storing column data in other MMDB servers.

2.1 About MonetDB

MonetDB is a typical DSM MMDB system, each column is stored independently in binary data structure named BAT with OID and data either on disk or in memory. BATs of same table have the same OID and OID value of same tuple is equal. For traditional SPJ operations, select operation applies only on relative BATs, project operation is efficient because only selected BATs are accessed, join operation occurs between two BATs with data matching firstly to get matched tuple OID set and then joining other output BATs with OID set.

Shown as figure 2, a SQL statement is divided into multiple BAT algebra operations, the operands and results are both BAT structure. SPJ operations can be divided into set of BAT operations. OLAP queries include many join operations, where clauses and aggregate operations, only small percent of fields are accessed during SQL statements executing and the results are always with small amount of fields and tuples. If SQL statements include low selectivity clauses and many output fields or SQL statements involve set operations, DSM MMDB is not superior to NSM MMDB. In our research work, we focus on OLAP applications which are suitable to MonetDB.

MonetDB supports two different kind of clients, one is SQL client, the other is MAL(MonetDB Assembly Language) client. There are two set of query processing modules for the two clients, MAL module acts as back end processing components in

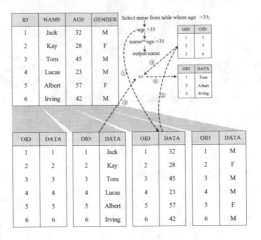

Fig. 2. BAT operations

MonetDB. A SQL statement can be divided into a set of independently processed MAL commands. It is difficult for us to directly re-writing SQL engine module, so we carry out our project with extending MAL modules.

2.2 Memory Hierarchy

We extend memory hierarchy with additional memory layer of "netMemory" which is built on high speed network for multiple MMDB nodes as figure 3 shows. Fibre optic technique supports high speed network even to Tb/s, while data access speed of disk is limited by disk rotating speed of around 100MB/s and although RAID 0 technique can improve data accessing speed for several times but proved to be less reliable for enterprise applications. The transmission speed of fiber optic technique is much higher than disk. Due to data accessing bottleneck of disk, the gap will become even larger in the future.

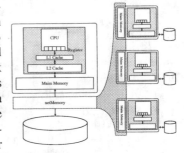

Fig. 3. netMemory

In our prototype system ScaMMDB, multiple MMDB server nodes have symmetrical or dissymmetrical computing power or memory capacity. A global data dictionary is replicated in each node for data distribution. Every node can independently response users' requirements and each BAT operation is transformed into local BAT operation or remote BAT operation according to global data dictionary, BAT can be replicated node-to-node by extended data accessing API, where clauses can be pushed down to BAT resident node for computing with only small result BAT returning back to requiring node. netMemory acts not only as dumb virtual data storage layer over low speed disk but also as column processing agent with extended computation power.

3 Architecture of ScaMMDB

ScaMMDB works in scalable application requirement scenarios based on shared-nothing architecture to obtain scalability of system capacity[2-4], users may meet such problems, for examples, scenario A: Data volume in DW exceeds system memory capacity, MMDB system is running with poor performance with a lot of I/Os. High-end server with large memory capacity is beyond budget, but there are several servers with total memory capacities satisfying memory requirements, can system enlarge memory capacity by assembling several low-end servers to establish a MMDB cluster and each node can response users' requirements independently? Scenario B: some servers may join system to enhance system capacity, some servers may leave system for some reasons, can system still work without re-install? Scenario C: Unfortunately, some servers are crashed suddenly; can system keep working without data lost? The target of project is to develop a shared-nothing MMDB cluster with scalability of memory storage and processing power, replication mechanism is considered for future work to improve reliability during m of N nodes fault. In ScaMMDB, our main targets are:

- ◆ Supporting dynamic memory sharing mechanism with remote BAT storing and accessing
- ◆ Supporting SQL statements automatically transforming into MAL commands with remote BAT accessing operations

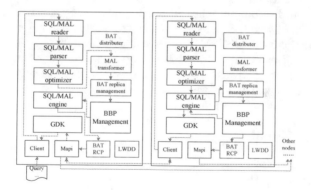

Fig. 4. Overall system architecture of ScaMMDB

Figure 4 shows the architecture of ScaMMDB, boxes with dashed line denote the extended modules. The functions of extended modules are as follows:

- ◆ LWDD(Light Weight Data Dictionary): storing distribution of columns in database, including meta data and nodeId of each column. LWDD is global maintained and synchronized when updating.
- ◆ BAT distributer: storing column data in different nodes according to LWDD.
- ◆ BAT RCP(Remote Call Processing): interface of remote BAT accessing. This module supports remotely calling BAT processing functions

in real BAT storage servers. BAT RCP module enables BAT processing on resident locations.

♦ MAL transformer: transforming formal MAL commands into ExMAL
commands with BAT RCP functions.

♦ BAT replica management: managing BAT creating, transferring, replicating, sharing and life cycle of remote BATs or remote result BATs.

When new node entering system to allow scale-up of memory capacity, we can use
two kind of BAT re-distributing policy to manage total memory resources. One is
incremental re-distributing policy. According to statistic information of BAT accessing frequency and workload of current system, migrating part correlative BAT from
heavy workload nodes to the new member node, LWDD in each node is updated to
record new memory resources. Another policy is to make global re-configuration with
additional memory capacity of new node. BAT distributer module analyzes meta
information of each node and tables to generate an optimized column data distribution
policy. Then re-distributing all the BATs in system, LWDD in every node is updated
subsequently.

When a node is going to leave system, BATs in memory are transferred to certain
nodes with free memory capacity or temporarily migrated to nearby nodes with some
low frequently accessed BATs storing on disk. At the same time, LWDD is refreshed
and broadcasted to all the nodes. After BATs are removed from current node, this
node can quit from system and re-distributing procedure can be called by system to
optimize query processing.

Comparing with distribution policy in SN parallel NSM databases, column(BAT)
is complete data unit for BAT algebra and can be replicated in other nodes without re-
distributing global data. In a NSM SN based parallel database system, distribution
policy such as hash or range distribution policy is global and when the amount of
servers is changed, all the distributed data has to be re-organized according new hash
or range policy based changing server amount. ScaMMDB support smoothly updating
system capacity with non-stop service.

4   Core-Modules of ScaMMDB

4.1   Optimization of BATs Distribution

Data distribution leads to transmission cost whatever in NSM model or DSM model
[8-10] and none of distribution policies can satisfy all queries with locally processing. In
parallel database, tables are distributed into multiple servers based on hash or range
distribution policies to reduce data transmission for queries. In DSM database, there
are many high selectivity clauses which can generate very small result data in column
operations. If we can generate co-operation column group in different servers then
only small result data need to be transferred between servers. In a typical OLAP application such as TPC-H database, types of queries are relatively stable, so analysis of
queries can provide meaningful information for data partition.

Our column distribution policy follows two rules, one is load balance, and the
other is co-operation column clustering. According to our performance test results, we
make BAT distribution plan with total size less than 75% of memory capacity to get

better performance, we also mix high frequency accessing BATs and low frequency BATs together to get better load balance for system. We optimize column distribution policy by clustering co-operation columns, such as join, group-by, order-by columns together to reduce column transmission cost.

We form a matrix for TPC-H queries analysis, Attribute Distance (AD) of attribute A_i and A_j can be defined as follows:

$$
\text{Dist}_{ij} = \begin{cases} 1 & A_i \text{ and } A_j \text{ have no affinity and } i \neq j \\ (\frac{1}{2})^n & A_i \text{ and } A_j \text{ have n times affinity and } i \neq j \\ 0 & i = j \end{cases} \tag{1}
$$

From TPC-H testing query Q1 to Q22, we divide each query into some field affinity groups, for example, "select A,B,C,SUM(D) from T1,T2 where T1.E=T2.F group by A,B,C" can be divided into two affinity groups, A,B,C,D form output affinity group and group-by affinity group, E and F form join affinity group. Each affinity group is then transformed into field pairs to identify relativity of each other. After establishing Attribute Distance matrix, clustering algorithm is applied to the AD matrix to get N clusters(N denotes the amount of nodes). Adjusting procedure then is called to adjust memory capacity based on number of clusters, memory capacity of each node, column accessing frequency, etc. Finally, column distribution policy generates LWDD for system and call distributer module to physically distribute BATs. Time cost of memory resident BAT processing is very small, so time cost of BAT transmission on Gbps NIC(Network Interface Card) influences query processing remarkably. With support of higher speed NICs of 10Gbps or Tbps, the time cost percent of transmission can be much smaller. Details can be referenced from another research paper of "A Data Distribution Strategy of Scalable Main-Memory Database".

4.2 Generation of ExMAL Commands

ExMAL represents Extended MAL which re-writes MAL commands into distributed MAL commands. The core function of transformation of MAL is to envelop common MAL command with remote calling function according to the location where BAT is stored. For example, "G_3:=algebra.join(G_1[:1],G_2[:2]);", G_1 is stored in node 1 and G_2 in node 2, we compare the sizes of G_1 and G_2, then select the larger BAT location to be operation locality and transferring another BAT to operation locality, the result BAT G_3 is generated in operation locality node.

We extend the system APIs to support remote BAT exchanging and shared BAT accessing between different threads. A set of transformation rules is created to enable original MAL commands to be re-written into ExMAL commands. Details can be referenced in our research work in NDBC 2008.

In MonetDB, a remote connection generates a processing thread. We use remote call procedure to control BAT processing among multi-node, the query responding node acts as temporary mediator node to generate remote call MAL commands. MAL commands for shared BAT processing through BBP(BAT Buffer Pool) are also generated by Gen-ExMAL algorithm. Figure 6 shows that ExMAL procedure includes remote BAT processing commands, connection management commands, shared BAT accessing commands and other commands such as temporary BATs management command.

Algorithm. GenExMAL(SQL,LWDD, rwRule)

```
1:    Transform SQL command into MAL commands MALset;
2:    Initiate system scenarios and connections;
3:    for each mal in MALset do
4:        ParsingMAL(mal, result, function, operandSet);
5:        executeNode=lowcostNode(operandSet);
6:        for all bat in operandSet do
7:            conn=seleConn(executeNode,bat);
8:            addToExMALsetRemoteGetOrPutBatMal(conn,bat);
9:        end for
10:       Exmal=matchRule(mal, LWDD, rwRule);
11:       addToExMALset(Exmal);
12:   end for
13:   free temporary BATs and connections;
```

GenExMAL algorithm can be seen as a naïve algorithm that directly translate MAL commands into remote call procedures. The same BAT operation may be called by different nodes in different threads in the same node as figure 5 shows, BAT operation is invoked privately for several times and the result must be stored in BBP for shared accessing. This mechanism increases complexity of MAL transformation. In our future work, we are going to develop a public BAT accessing interface for remote accessing and integrate BAT transmission and shared accessing functions. The unique public interface can simplify remote BAT accessing procedure just like web service in web function integration.

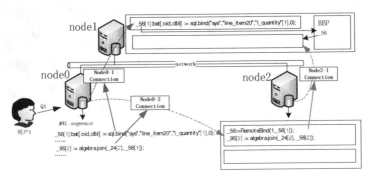

Fig. 5. Remote BAT processing

4.3 BAT Transmission Module

In our experiments, BAT transmission is a costly procedure for ExMAL commands. Optimizing BAT distribution policy can reduce BAT transmission cost but there are still some big BATs need to be transferred between nodes. Under our experiment environment, the data accessing speed of disk and network is about 70MB/s:100MB/s, loading BAT from netMemory is faster than from disk. A complete BAT transmission procedure includes three steps: read column data from BAT into sending buffer, transfer data between nodes, clone a new BAT with column data of receiving buffer.

The most important consideration of this module is efficiency. Firstly, we develop BAT transmission module bases on system API of reading and writing BAT storage unit. There are millions of loops of reading or writing operations during one BAT transmission operation(1GB TPC-H database has 6,000,000 tuples in fact table), the time cost of reading millions of data units into sending buffer and writing millions of data units into cloned BAT from receiving buffer are much higher than column transmission itself. Finally, we optimize BAT transmission API with BAT memory mapping which uses source BAT storage area as sending buffer and target BAT storage area as receiving buffer, so time costs of BAT decomposing and re-constructing can be saved. The fixed length type BAT can employ BAT memory mapping API but variable length type BAT stores offset addresses in BAT storage area and all real data in another memory area which has to transfer two independent memory contents and re-assign offset addresses in target BAT.

1Gb/s NIC can provide about 100MB/s data transmission rate. Higher performance NIC such as 10Gb/s or Tb/s NIC can provide even higher data transmission rate. netMemory based on large amount of inexpensive servers can improve the scalability of main memory databases. Combined with replication policy for reliability, Redundant Array of Independent Memory Network(RAIMN) can also developed for higher memory capacity and reliable requirements application backgrounds. In OLAP application scenario, TB level data warehouse exceeds the processing capacity of single MMDB server, RAIMN architecture can combine many inexpensive servers as a large virtual memory pool for mass data processing. Based on network technique and netMemory architecture, MMDB cluster, MMDB replication service and RAIMN facing very large database MMDB application will gradually become practical.

5   Experiments

In order to get a credible experiment results, we perform our test with TPC-H database and testing queries. We perform our test on 3 HP Intel Itanium-64 servers with two 1.6G CPUs running Red Hat Enterprise Linux ES release 4 (kernel version 2.6.9) with 2GB RAM, 2GB Swap and 80GB hard disk.

A. Client and main functions of ScaMMDB
Figure 6 shows ScaMMDB interface. There are three nodes in experiment system; we can choose any node as working node. (a) shows the query interface of query execution, it is same as single server. (b) displays the analysis information of ExMALs, we can get information of time cost and time cost rate of each ExMAL command to analyze how to improve performance.

Comparing with original MAL commands, system adds connection maintenance commands, remote BAT accessing command and envelops remote call BAT operations in ExMAL commands.

B. Performance of BAT transmission

In figure 7, we compare the transmission performance of different type data of BATs. Ideal network speed of Gbps NIC is computed as 128MB/s, but influenced by additional network packets data, main board bus speed and other factors, the real data transmission speed is less than 128MB/s. In our experiment, transmission speed is computed as: $T(MB/s)=sizeofBAT(MB)/timeCost(s)$.

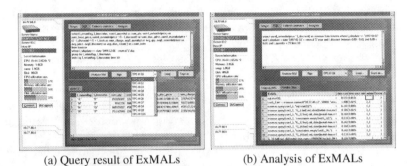

(a) Query result of ExMALs   (b) Analysis of ExMALs

Fig. 6. Overall time cost of delta table merging algorithms

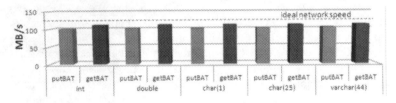

Fig. 7. BAT transmission performance

BAT transmission includes two types of getBAT and putBAT which enable node can "pull" or "push" BAT from or to other node. The speed of getBAT is a little more than putBAT because of no need of block operation when generating target BAT. The average speed of BAT transmission is about 110MB/s which is very close to ideal network speed and more than data accessing speed from disk.

C. Performance of TPC-H queries in ScaMMDB

We select TPC-H testing queries of Q1,Q6,Q10,Q13 as targets. According to given LWDD, different queries have different BAT copy time cost. For example in Q6(node 2), the cost of BAT copy takes more than 60 percents of total time cost. If we use high speed NIC of 10Gbps or Tbps, the BAT copy cost will be remarkably reduced as figure 8(a) shows. In ScaMMDB, remote MAL call take additional cost than local MAL command of average 0.89% more time in 4GB testing data and it will be smaller as volume of database grows. With modern network technique, ScaMMDB can extend data and processing capabilities with acceptable network latency.

There are three main types of remote BAT accessing policies. The "M" mode use other node as secondary memory, replicating needed remote BAT first and then processing query with local BAT replicas; the "N" mode means that when a MAL

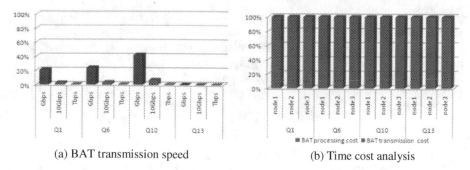

(a) BAT transmission speed   (b) Time cost analysis

Fig. 8. BAT transmission time cost rate

command has multiple parameters which have different locations, system chooses working node which has the maximum number of BATs or local node when numbers are equal; the "S" mode uses total transmission size instead of numbers as parameter. Both "N" and "S" modes are local optimizing policies based on current ExMAL command, global optimizing policy is more complex and in our future research works.

"S" mode has nearly equal time cost of BAT transmission but "N" mode has different cost in different node e.g. Q6 in node 1 and node 2. Figure 8(b) shows the time cost of testing queries in each node, we can see that with low speed NIC configuration, time cost of BAT transmission effects performance greatly and BAT distribution policy is important for performance of system.

6 Conclusions and Future Work

Comparing with typical cluster system e.g. Oracle RAC or Shared-Nothing parallel systems, the key point of ScaMMDB is the column based distributing policy and column-centered processing. In RAC system, single database is accessed by multiple instances, both disks and memory are shared by multiple instances. But in main memory database systems, data is memory resident and memory capacity is limited, the key problems are how to distribute data among multiple nodes and how to reduce redundant data replicas when data in other node. Most SN parallel DBMSs are horizontally partitioned with row-store tuples. In Data Warehouse, fact tables have millions of tuples and multiple dimensional attributes which will join with multiple dimensional tables. It is difficult to select suitable horizontal partition policy to avoid data transferring when fact table joins with multiple tables. For column-store MMDB, column is the basis unit for distributing and not only column itself but also processing upon column can be shared among multiple nodes.

We develop prototype of ScaMMDB to improve scalability of MMDB in modern OLAP application scenarios. According to techniques of modern hardware of large memory and high speed network, netMemory is presented to extend memory hierarchy for dynamical system re-organization in this paper. MMDB based on netMemory can provide scalable memory capacity for OLAP applications, and we extend MonetDB to support query processing over multi-node system. ScaMMDB provides a mechanism to solve mass data processing with high performance MMDB. In our present prototype, we didn't consider redundant BAT mechanism because of memory capacity, most time cost is BAT transmission cost, appropriate redundant BAT mechanism can improve performance further. We will continue our research on optimizing ExMAL commands and distribution policy. MMDB cluster is another research topic in our future work.

References

1. http://www.wintercorp.com/VLDB/2005_TopTen_Survey/
   TopTenWinners_2005.asp
2. Han, W.-S., et al.: Progressive optimization in a shared-nothing parallel database. In: Proc. SIGMOD, Beijing,China, pp. 809–820 (2007)
3. Antunes, R., Furtado, P.: Hardware Capacity Evaluation in Shared-Nothing Data Warehouses. In: Parallel and Distributed Processing Symposium, IPDPS, pp. 1–6 (2007)
4. Bamha, M., Hains, G.: A skew-insensitive algorithm for join and multi-join operations on shared nothing machines. In: Ibrahim, M., Küng, J., Revell, N. (eds.) DEXA 2000. LNCS, vol. 1873, pp. 644–653. Springer, Heidelberg (2000)
5. Abadi, D.J., Madden, S.R., Hachem, N.: Column-Stores vs. Row-Stores: How Different Are They Really? In: Proc. SIGMOD, Vancouver, Canada (2008)
6. Stonebraker, M., et al.: C-Store: A Column-oriented DBMS. In: Proc. VLDB, Trondheim, Norway, pp. 553–564 (2005)
7. Zukowski, M., Nes, N., Boncz, P.A.: DSM vs. NSM: CPU Performance Tradeoffs in Block-Oriented Query Processing. In: Proc. the International Workshop on Data Management on New Hardware (DaMoN), Vancouver, Canada (2008)
8. Ghandeharizadeh, S., DeWitt, D.: Hybrid-range partitioning strategy: a new declustering strategy for multiprocessor database machines. In: Proc. VLDB, Brisbane, pp. 481–492 (1990)
9. Jianzhong, L., Srivastava, J., Rotem, D.: CMD: A multi-dimensional declustering mothod for parallel database system. In: Proc. VLDB, VanCouver, pp. 3–14 (1992)
10. Ghandeharizadeh, S., DeWitt, D.J.: A performance analysis of alternative multi-attribute declustering strategies. In: Proc. SIGMOD, San Diego, California, pp. 29–38 (1992)

A Data Distribution Strategy for Scalable Main-Memory Database

Yunkui Huang[1,2], YanSong Zhang[1,2], XiaoDong Ji[1,2], ZhanWei Wang[1,2],
and Shan Wang[1,2]

[1] Key Lab of Data Engineering and Knowledge Engineering, Ministry of Education, China
[2] School of Information Renmin University of China, China
{hyk,zhangys_ruc,jixiaodong,mayjojo,swang}@ruc.edu.cn

Abstract. Main-Memory Database (MMDB) System is more superior in less response times and higher transaction throughputs than traditional Disk-Resident Database (DRDB) System. But the high performance of MMDB depends on the single server's main memory capacity, which is restricted by hardware technologies and operating system. In order to resolve the contradiction between requirements of high performance and limited memory resource, we propose a scalable Main-Memory database system ScaMMDB which distributes data and operations to several nodes and makes good use of every node's resource. In this paper we'll present the architecture of ScaMMDB and discuss a data distribution strategy based on statistics and clustering. We evaluate our system and data distribution strategy by comparing with others. The results show that our strategy performs effectively and can improve the performance of ScaMMDB.

Keywords: MMDB, ScaMMDB , Data Distribution Strategy.

1 Introduction

In the recent years, with the rapid development of data warehouse, data mining and other OLAP applications, traditional Disk-resident Database (DRDB) System can not meet the requirement of high performance. More and more researchers have focused on Main-Memory database (MMDB) system. The key difference between DRDB and MMDB is that the primary copy of data in MMDB lives permanently in memory, data can be accessed directly in memory and little time is spent on I/O which is the bottleneck of DRDB. Thus MMDB can provide much better response times and transaction throughputs.

But the high performance of MMDB depends on the single server's memory capacity, which is restricted by hardware technologies and operating system. For example, in OLAP applications, data volume often arrives at several TB [1], but no ordinary single sever can provide TB memory now. After carefully researching and analyzing, we can draw a conclusion that the performance of MMDB will drop rapidly when the data volume exceeds 75% of memory capacity, and it will not work any more when it reaches 200%. We'll present strong evidence in section 4. Thus

L. Chen et al. (Eds.): APWeb and WAIM 2009, LNCS 5731, pp. 13–24, 2009.

there's a huge contradiction between requirement of high performance and limitation of single server's memory capacity. In order to solve this contradiction, we can't move back to traditional DRDB, since network transport will definitely be much faster than disk I/O in the future, we implement a scalable Main-Memory Database System by combining several nodes together with high speed network, it is called ScaMMDB, which mainly focuses on improving the performance of OLAP and other Query-Intensive applications.

ScaMMDB is a system based on column stored model, having Shared Nothing (SN) architecture. The most important advantages of ScaMMDB are breaking up the restriction of single node's memory capacity and distributing the data and operations to several nodes. However, the cost of transporting between nodes is an additional negative factor for high performance. So it is extremely important to find an effective way to minimize the cost of accessing data items during transaction processing.

In fact, transport cost totally depends on the data distribution strategy, which is really the key point of improving performance. In this paper, we present a data distribution strategy based on statistics and clustering. In the case of this strategy, transaction processing cost is minimized by increasing the local processing of transactions (at a site) as well as by reducing the amount of accesses to data items that are remote. We analyze every executed query, calculate the distance of each two attributes, and then clustering attributes based on the attribute distance matrix, the result is that attributes in one cluster are correlative as much as possible and attributes between clusters are irrelative. We distribute the data in one cluster to one single node of ScaMMDB, and also take the processing capability and memory capacity of every node into account. We'll evaluate this strategy by comparing with typical vertical partitioning methods in distributed database latter.

In the past decades, with the depreciating of random access memory and widely using of the 64 bit CPU, more and more companies and researchers have focused on MMDB, and a few of prototype systems and commercial systems have been developed. Such as Dali[2], TimesTen [3] and MonetDB [4, 5]. On the one hand, researchers primarily focus on the MMDB itself. Such as the optimization of Cache[6], index technology [7], backup and restore etc. few researchers care about the issues of Parallel and Distributed Main-Memory Database System, only a few discuss some commit protocol[8] and other key technologies[9]. No integrated scalable MMDB has been implemented until now.

On the other hand, according to distributed DRDB, there are primarily three kinds of data partition methods. Horizontal partition, Vertical partition and mixed partition. Most of the Distributed DRDB adopts Horizontal partition method because of their row stored model. But ScaMMDB adopts the column stored model, so we mainly discuss some vertical partitioning algorithms here. In the past decades, several vertical partitioning algorithms have been proposed. Hoer and Severance [10] measure the affinity between pairs of attributes and try to cluster attributes according to their pair wise affinity by using the bond energy algorithm (BEA) [11]. Navathe and Ra [12] construct a graph-based algorithm to the vertical partitioning problem. where the heuristics used includes an intuitive objective function which is not explicitly quantified. Jeyakumar Muthuraj [13] summarizes the formal approach of vertical partition problem in distributed database design. We will evaluate our algorithm by comparing with BEA latter.

The rest of the paper is organized as follows. In section 2, we introduce the general architecture of ScaMMDB. We mainly discuss the three stages of our data distribution strategy in section 3. Our experimental results will be reported in section 4, and in section 5, we make a conclusion.

2   System Architecture

ScaMMDB is a scalable MMDB prototype system which includes several nodes, complex queries can be executed concurrently. The architecture of ScaMMDB (Fig.1 shows) is based on Shared Nothing (SN) structure. Every node has its own CPU, Memory, Disk and other hardware. Nodes are connected by a gigabit network. Every two nodes are equal, and none is master or slave. We optimize the storage architecture and add one level of NetMemory, which is the memory resource of other nodes. On every node, there is a MMDB based on the open source MMDB MonetDB [5] which adopts the column stored model. Each column is stored as a Binary Association Table (BAT). BAT is the smallest object of operation. Query is transformed into MonetDB Assembly Language (MAL). MAL is a kind of execution plan, also a sequence of operations. We expand MAL by implementing an Extended Parallel MAL (EPM) interface, and then adding the operations of visiting, transporting and rebuilding the Remote BAT, which enable every node to participate in the data processing.

Data Distribution Strategy (DDS) is the foundation of ScaMMDB. Different DDS will give rise to data being distributed in different nodes, and the performance will be different too. Light Weighted Data Dictionary (LWDD) is the representation of DDS, and also is the foundation of query executed. Every node has its own LWDD, which stores the information of every attribute and every node, such as attribute name, BAT size, node ID that it belongs to, IP address and memory capacity of every node. LWDD is updated simultaneously when the DDS is changed.

We can propose query request on any node at any time. After query reading, parsing, rewriting and optimizing stage, query is transformed into MAL clauses, then each clause is distributed to the relevant node according to the involved attributes' location information which can be found in LWDD, and the results are finally gathered to the query requested node. So data transport is necessary.

The cost model of ScaMMDB must be as follows:

$$T = T_{Memory} + T_{NetMemory} = T_{CPU} + T_{TLB} + T_{L1} + T_{L2} + T_{L3} + T_{NetMemory} \qquad (1)$$

Where: T_{CPU} is the CPU computing time. T_{TLB} is the address translating time. Address translation is carried on in Translation Look-aside Buffer (TLB). T_{Li} is L_i level cash missing time. $T_{NetMemory}$ is the transporting time among nodes.

Network transport speed is faster than disk I/O speed now, and with the rapid development of network technology, the gap will be larger and larger. But now, In ScaMMDB, transport cost is still the bottleneck. How to decrease $T_{NetMemory}$ makes a great impact on performance. In fact, $T_{NetMemory}$ is decided by the data distribution and network transporting speed. Find an effective data distribution strategy is the key

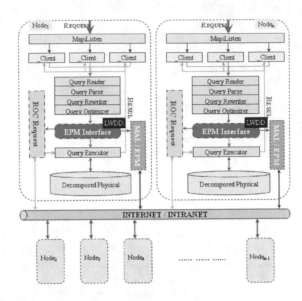

Fig. 1. Architecture of ScaMMDB

point of improving performance of ScaMMDB. The following part of this paper will mainly discuss a data distribution strategy based on statistics and clustering.

3   Data Distribution Strategy

3.1   Overview

Our Data Distribution Strategy (DSS) is a statistical information based algorithm. Generally speaking, the system must experience a period of learning, accumulating attributes distance information, and then redistributing the data based on this information. That is because of the principle of 80/20. There is a serious "Skew" phenomenon when attributes are visited by queries in practical applications. That means some attributes are visited by every queries for several times, but some attributes never be visited by a query. So we accumulate the attribute affinity information based on executed queries in the learning period. New queries will also comply with this information after learning period.

The data distribution strategy can be divided into three stages. Firstly, generating Attribute Distance Matrix (ADM) based on user input or statistical information. Secondly, clustering attributes into several clusters, and attributes in each cluster are distributed into one node of ScaMMDB. Finally, compute the total data volume of every single node, if the total data volume excesses 75% of node's memory capacity, an adjustment must be performed. The following part of this paper will discuss details.

3.2   Generate Attribute Distance Matrix

We take estimated total data volume S and number of nodes M into account at first. If S is smaller than the minimum memory capacity of all nodes, just copy the data to every node in order to get high performance and parallelism. If S is larger than total memory capacity of all nodes, ScaMMDB can not deal with this situation, more nodes are needed. On the other hand, if the ratio of attributes number to node number is too small (for example, less than 5), that means there are too many nodes and few attributes, transport cost will be too high. User need to get rid of a few nodes or do some data duplications between nodes. The following part of this paper mainly discuss the normal situation not these extreme cases.

The attribute affinity can be defined preliminarily by user or we can analyze the executed query. Attribute affinity is closely related to data storage model, query structure and executing methods. For ScaMMDB, on one hand, user can define which attributes are affinity and the distance of each two attributes, that's to say, user can define Attribute Distance Matrix (ADM) at first as input parameter. ADM is a symmetric matrix whose dimension is the same as number of attributes. On the other hand, we analyze every executed query at query parsing stage, and store the attribute distance as ADM based on some principles simultaneously. If each two attributes have one time affinity, we update the corresponding value in ADM.

In ScaMMDB, attributes have one time affinity is defined as follows:

1. Putting "select" attributes, "group by" attributes and "order by" attributes into one set, any two attributes in this set have one time affinity. Because the "select" attributes must be outputted according to "group by" attributes and "order by" attributes on the requested node.

2. Putting all the attributes involved in "and" clause into one set, there may be join operations, comparison operations and arithmetic operation.

These principles may only be suitable for ScaMMDB because of its special storage model, query executing methods and Monet Accessing Language (MAL). But there must be an attribute affinity principle being suitable for the other systems.

Attribute Distance (AD) of attribute A_i and A_j can be defined as follows. Definitely, weights of different attribute size should be different. We must take attribute size S_i and S_j into account, take average size of attributes S_{avg} as a const factor.

$$
\text{Dist}_{ij} = \begin{cases} 1 & A_i \text{ and } A_j \text{ have no affinity and } i \neq j \\ \left(\dfrac{S_{avg}}{S_i + S_j + S_{avg}}\right)^n & A_i \text{ and } A_j \text{ have n times affinity and } i \neq j \\ 0 & i = j \end{cases} \qquad (2)
$$

In the process of parsing every query Q, the attribute distance will be computed, and ADM will be updated according to Algorithm 1. If user has already defined ADM at first, we do not need to parse queries and get affinity information. The ADM will be stored permanently unless user sends a reset command.

Algorithm 1. GenerateADM (Q, is_defined, need_init)

Require: Q: the input query; is_defined: whether ADM has defined by user;

need_init: whether ScaMMDB needs initialization.

1: if is_defined == TRUE then
2:       return
3: else
4:       if need_init == TRUE then
5:             InitADM() /*traverse all tables, generate a initial ADM dist[N][N] for N
             attributes. dist[i][j] = 1 only if $i \neq j$, and if i = j, dist[i][j] = 0. */

6:       else
7:             read dist[N][N] from disk
8:       end if
9:       while Q != NULL do
10:            Parse(Q)   /*parse the query. */
11:            find all <i, j> that makes attribute Ai and Aj have one time affinity
12:            for each <i, j> do
13:                  dist[i][j] = dist[i][j] * 0.5
14:            end for
15:       end while
16:       write dist[N][N] to disk
17: end if

If ScaMMDB runs at the first time and no ADM is defined by user, a default initial ADM will be created. In the matrix dist[N][N], the distance of each two different attributes is 1, and cater-corner value of dist[N][N] is 0. ADM will be written to disk after query executed completely. Next step will be attributes clustering.

3.3   Attributes Clustering

There are so many clustering methods in data mining area, such as partitioning method, hierarchical method, density-based method, grid-based method and model-based method. Each method has its own scope of application. There are not too many attributes in practical applications, especially for OLAP application. However, a lot of complex queries bring on "Skew" phenomenon for visiting attributes. That means so many values of the Attribute Distance Matrix (ADM) are 1. Hierarchical clustering method is simple, easy to implement and more suitable than other methods.

There are a set of N attributes to be clustered, an N*N distance matrix dist[N][N], and M nodes in ScaMMDB. Starting by assigning each attribute to a cluster, so that if you have N attributes, you now have N clusters, each contains just one attribute. Then find the closest (most similar) pair of clusters and merge them into a single cluster, delete the rows and columns corresponding to these two clusters, so that now you have one cluster less, then compute distances between the new cluster and each of the old clusters. Repeat until all attributes are clustered into M clusters.

For computing distance between clusters, there are many different ways of defining distance D(r, s) between cluster r and cluster s.

Single linkage clustering. The defining feature of the method is that distance between groups is defined as the distance between the closest pair of objects, where only pairs consisting of one object from each group are considered. In the single linkage method, D(r, s) is computed as

$$D(r,s) = \text{Min } \{Dist_{ij}: \text{Where object i is in cluster r and object j is cluster s }\} \qquad (3)$$

Complete linkage clustering. Also called farthest neighbor, clustering method is the opposite of single linkage. Distance between groups is now defined as the distance between the most distant pair of objects, one from each group. D(r,s) is computed as

$$D(r,s) = \text{Max } \{Dist_{ij}: \text{Where object i is in cluster r and object j is cluster s }\} \qquad (4)$$

Average linkage clustering. Here the distance between two clusters is defined as the average of distances between all pairs of objects, where each pair is made up of one object from each group. D(r,s) is computed as

$$D(r,s) = \frac{T_{rs}}{N_r * N_s} \qquad (5)$$

Where T_{rs} is the sum of all pairwise distances between cluster r and cluster s. N_r and N_s are the sizes of the clusters r and s respectively. At each stage of hierarchical clustering, the clusters r and s, for which D(r,s) is the minimum, are merged.

Average group linkage. Groups once formed are represented by their mean values for each variable, that is, their mean vector and inter-group distance is now defined in terms of distance between two such mean vectors. The two clusters r and s are merged such that the average pairwise distance within the newly formed cluster is minimum. Suppose we label the new cluster formed by merging clusters r and s as t. The distance between clusters r and s is computed as

$$D(r,s) = \text{Average } \{Dist_{ij}: \text{Where observations i and j are in cluster t.}\} \qquad (6)$$

Ward's method. Ward proposed a clustering procedure seeking to form the partitions P_k, P_{k-1}, ..., P_1 in a manner that minimizes the loss associated with each grouping and to quantifies that loss in readily interpretable form. Information loss is defined by Ward in terms of an error sum-of-squares (ESS) criterion. ESS is defined as the following

$$ESS = \sum_{K=1}^{K} \sum_{xi \in C_k} \sum_{j=1}^{p} (x_{ij} - \overline{x_{kj}})^2 \qquad (7)$$

With the cluster mean $\overline{x_{kj}} = \frac{1}{n_k} \sum_{xi \in C_k} x_{ij}$, where x_{ij} denotes the value for the i-th individual in the j-cluster, k is the total number of clusters at each stage, and n_j is the number of individuals in the j-th cluster.

We will evaluate the result of clustering in section 4, especially these different ways of distance defining methods between clusters. According to the experiments, Average group linkage method is the most useful methods for ScaMMDB. After all attributes are clustered into M clusters, an adjustment must be needed.

3.4   Adjustment

All attributes have been distributed into M nodes, attributes in one node have high affinity and attributes between nodes have low one. But the processing ability of each node and available memory is limited. So the result of clustering must be adjusted in order to make good use of every node in ScaMMDB.

Let's make a definition:

$$SC_i = \frac{DS_i}{MC_i} * 100\% \qquad (8)$$

Where: DS_i is the data volume of node i, MC_i is the Memory capacity of node i

We can get a high performance on a single node only if the SC is under 75%. So Adjustment is needed on the nodes whose SC is over 75% and redistribute some columns to the other nodes which have enough memory capacity and high affinity with. Local node is the node of results exhibiting. We should distribute the cluster which has the most "select" attributes to local node. The algorithm is as this:

Algorithm 2. Adjustment(Si)

Require: S_i: size of attribute i
1: compute SC_j for every node j
2: for each node N_j whose $SC_j > 75\%$ do
3:      while $SC_j > 75\%$ do
4:          $s = \min\{S_i \mid$ attribute i in node $N_j\}$ /* find the smallest data volume of attribute in node N_j*/
5:          find node N_k that $SC_k < 75\%$ after add attribute i and each attribute in N_k have smallest sum distances with attribute i. no N_k is found, report errors and more nodes needed.
6:          transform attribute i from N_j to N_k and compute the SC_j
7:      end while
8:      find the node k which has the most "select" attributes;
9:      if $k \neq local\_node$
10:         exchange the attributes between node k and local node
11: end for

After analyzing queries, generating ADM, clustering and adjusting, attributes are distributed into several nodes based on our data distributed strategy. ScaMMDB will generate Light Weighted Data Dictionary (LWDD) according to the attribution distribution. LWDD is the foundation of remote column operation and query execution. At last, we will evaluate this strategy by the experimental results.

4   Experiments

Our experiments have been designed to evaluate the foundational principle of ScaMMDB, the effect of different distance computing methods between clusters in our strategy and the performance between our strategy and other methods.

ScaMMDB involved in three nodes, each node has two Intel Itanium-64 1.6G CPU running Red Hat Enterprise Linux ES release 4 (kernel version 2.6.9) with 2GB RAM, 2GB Swap and 80GB hard disk. Nodes are connected by Gigabit networks. The baseline version of ScaMMDB is MonetDB Server v5.0.0. We adopt TPC Benchmark™ H [18] data sets as our test data sets and implement the TPCH test tool TPCH-Tester on our own.

Firstly, we carry out TPC Benchmark™ H based on an open source MMDB MonetDB, compute the TPC-H query processing power TPCH Power@Size, We primarily focus on Query-Intensive applications and leave refresh stream out of account. So we define Power@Size as follows:

$$Power@Size = \frac{3600 * SF}{\sqrt[22]{\prod_{i=1}^{i=22} QI(i,0)}} \tag{9}$$

Where: SF is the scale factors, $QI(i,0)$ is the timing interval, in seconds, of query Qi within the single query stream of the power test.

The high performance of MMDB (Fig.2 shows) is just limited to the situation that SC (rate of data volume to memory capacity) is below 75%, the performance falls sharply when SC is over 75%, and system will breakdown when SC arrives at 200%.

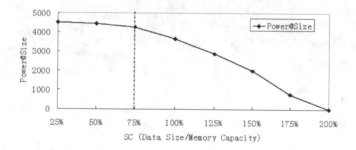

Fig. 2. Power@Size under different SC

So 75% of SC is the critical point of single node's high performance. This also provides an important and reasonable foundation for scalable Main-Memory database and data distribution strategy.

Secondly, in order to select the best suitable methods for clustering under the specific applications of ScaMMDB. We carry out experiments under the data volume of 3GB, and do comparison between different methods of computing distance between clusters. Unfortunately, clustering based on single linkage clustering, complete linkage clustering and average linkage clustering give rise to the data distribution seriously skew (Fig.3. shows), data volume in Node1 is 2.6GB of 3GB total, far exceeds its memory capacity. That illustrates the necessity of doing an adjustment.

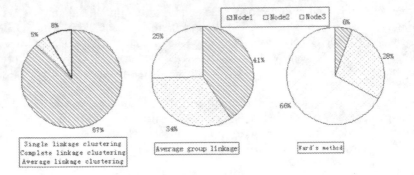

Fig. 3. Data distribution under different cluster distance computing methods

Obviously, single linkage clustering, complete linkage clustering and average linkage clustering are not suitable for this attribute relationships because of their terrible clustering results, and adjustment will not take effect either.

We compare the Average group linkage with Ward's method following an adjustment. For Statistic and Clustering Algorithm, based on the statistical information of queries Q1 to Q15 which are executed before on local node, we regenerate LWDD, executing all of the 22 queries under the newly generated LWDD and then computing the transmission quantity. Every node is equal, so queries requested on any node takes the same effect. The result (Fig.4 shows) demonstrates that Average group linkage is more suitable under these OLAP applications

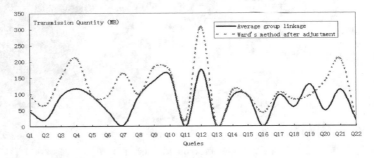

Fig. 4. Transmission Quantity under different cluster distance computing methods

Finally, we carry out experiments under the data volume of 3GB, and compare the results of different data distribution strategies. These are Round-robin, Bond Energy Algorithm (BEA) [13] and our Statistic and Clustering Algorithm. Round-robin is a fully equal distribution method, and BEA is one of the most common and famous algorithms in distributed database design. It generates a clustered affinity matrix (CAM) at first, and then does binary vertical partition recursively. But BEA doesn't take the query semantics and dealing ability of nodes into account. There is definitely an imbalance of data distribution. In Statistic and Clustering Algorithm, we adopt the Average group linkage method to compute the distance of different clusters.

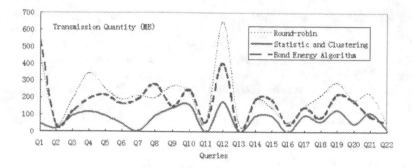

Fig. 5. Transmission Quantity under different data distribution strategy

We can see from the results (Fig.5 shows), Round-robin as a naïve strategy gives rise to the most transmission Quantity, and Bond Energy Algorithm also can not do better than the Statistic and Clustering Algorithm. Transmission quantity is close to zero for some queries, because the related attributes are already been clustered into one node.

5   Conclusion

In this paper, we proposed the architecture of scalable Main-Memory database system and primarily discuss a statistic and clustering based data distribution strategy. Unlike other methods, we generate the ADM (Attribute Distance Matrix) by analyzing the executed queries and also take every node's processing ability into account. We distribute data between nodes as equally as possible and every node plays a great role in the process of query executing. The evaluation experiments show that our strategy obtains an improvement of effectiveness over others.

In the future, there will be a complex, iterative and dynamic process of learning according to generate attribute distance matrix. We will do more work on designing the algorithms of query learning and LWDD rebuilding, Algorithms in machine learning may take effect. We will try to optimize our approach by combining the adjustment and clustering together, take attribute affinity and data volume into account simultaneously. Generating data dictionary will be an iterative processing in order to obtain an optimum data distribution. Our strategy is not the only one solution for ScaMMDB. This approach fits for ScaMMDB under OLAP does not mean it is suitable for every scalable Main-Memory database system under any applications. We will try to explore a general strategy under OLTP and other common applications.

References

1. Organizations and their vendors for their achievements in the 2005 TopTen Program,
   http://www.wintercorp.com/VLDB/2005_TopTen_Survey/
   TopTenWinners_2005.asp

2. Jagadish, H.V., Lieuwen, D., Rastogi, R., Silberschatz, A., Sudarshan, S.: Dalí a high performance main memory storage manager. In: Proceedings of the 20th International Conference on Very Large Data Bases (VLDB 1994), pp. 48–59 (1994)
3. Oracle TimesTen In-Memory Database Architectural Overview Release 6.0, http://www.oracle.com/database/timesten.html
4. Boncz, P.A.: Monet: A Next-Generation DBMS Kernel for Query-Intensive Applications. Ph.D. Thesis, Universiteit van Amsterdam, Amsterdam, The Netherlands (2002)
5. MonetDB, http://monetdb.cwi.nl
6. Manegold, S., Boncz, P., Nes, N.: Cache-conscious radix decluster projections. In: Proceedings of thirtieth International conference on Very Large Data Bases (VLDB 2004), pp. 684–695 (2004)
7. Luan, H., Du, X., Wang, S.: J + 2Tree: a new index structure in main memory Database Systems for Advanced Applications. In: Kotagiri, R., Radha Krishna, P., Mohania, M., Nantajeewarawat, E. (eds.) DASFAA 2007. LNCS, vol. 4443, pp. 386–397. Springer, Heidelberg (2007)
8. Lee, I., Yeom, H.Y., Park, T.: A New Approach for Distributed Main Memory Database System: A Causal Commit Protocol. IEICE Trans. Inf. & Syst. E87-D, 196–296 (2004)
9. Chung, S.M.: Parallel Main Memory Database System. Department of Computer Science and Engineering Wright State University (1992)
10. Antunes, R., Furtado, P.: Hardware Capacity Evaluation in Shared-Nothing Data Warehouses. In: IEEE International Parallel and Distributed Processing Symposium, pp. 1–6 (2007)
11. Han, W.-S., et al.: Progressive optimization in a shared-nothing parallel database. In: Proceedings of the 2007 ACM SIGMOD international conference on Management of data, pp. 809–820 (2007)
12. Hoer, J., Severance, D.: The Uses of Cluster Analysis in Physical Database Design. In: Proc. 1st International Conference on VLDB, pp. 69–86 (1975)
13. McCormick, W., Schweitzer, P., White, T.: Problem Decomposition and Data Reorganization by a Clustering technique Operations Research (1972)
14. Navathe, S., Ra, M.: Vertical Partitioning for Database Design: A Graphical Algorithm. ACM SIGMOD (1989)
15. Muthuraj, J.: A Formal Approach to the Vertical Partitioning Problem in Distributed Database Design. University of Florida (1992)
16. Han, J., Kamber, M.: Data Mining: Concepts and Techniques. Morgan Kaufmann Publisher, San Francisco (2000)
17. Jain, A.K., Murty, M.N., Flynn, P.J.: Data clustering: A survey. ACM Comput. Surv. 31, 264–323 (1999)
18. TPC Benchmark™ H, http://www.tpc.org
19. Johnson, S.C.: Hierarchical Clustering Schemes. Psychometrika 2, 241–254 (1967)
20. D'andrade, R.: U-Statistic Hierarchical Clustering. Psychometrika 4, 58–67 (1978)

Temporal Restriction Query Optimization for Event Stream Processing

Jiahong Liu, Quanyuan Wu, and Wei Liu

Institute of Network Technology & Information Security School of Computer,
National University of Defense Technology, Changsha 410073, China
kahon@nudt.edu.cn, wuquanyuan@nudt.edu.cn, liuwei2work@aol.com

Abstract. The trend that organizations are linking Service Oriented Architecture (SOA) efforts closely to real-time processes makes research and industrial community increasingly focus on the SOA and Event Stream Processing (ESP) connection. ESP needs to correlate multiple continuous events involved in complex temporal relationship and attribute logic relationship to more abstract complex events in richer semantic. Due to high speed arrival rate of events and vast volume of registered complex event queries, memory consumption and incremental event query evaluation demand a comprehensive dedicate event stream processing framework with low-latency and high scalability. In this paper, we study problems of query optimization for ESP, especially topics on temporal restriction query. We first propose a framework to integrate ESP features with business process management and monitor. We then describe a query plan-based approach to efficiently execute ESP queries. Our approach uses algebraic conversion to efficiently handle temporal restriction queries, which are a key component of complex event processing, and computes temporal relevance condition at compile time to obtain event relevance time for a given expression. We demonstrate the effectiveness of our approach through a performance evaluation of our prototype implementation.

1 Introduction

Organizations are re-architecting of existing business processes with Service Oriented Architecture (SOA) principals to provide integration and interoperability, meanwhile different industry sectors are using Event Stream Processing(ESP) for all critical business processes, thus pushing its span beyond financial and simulation applications. The trend that organizations are linking SOA efforts closely to real-time processes, makes research and industrial community increasingly focus on the SOA-ESP connection[1]. Many complex event queries in ESP need to detect interested events during intended time scopes[2]. In this paper, we study the query optimization problem in a SOA monitoring environment that consists of potentially a large number of distributed event sources. We specially focus on topics on temporal restriction query.

Temporal restriction is a key component for complex event query in ESP. It determines temporal constraints for target queries. Many works model temporal restriction as window query and temporal selection condition during query evaluation [2, 3]. In this paper we will leverage algebraic equivalence of event expression with temporal

L. Chen et al. (Eds.): APWeb and WAIM 2009, LNCS 5731, pp. 25–35, 2009.

restriction, then at compile time compute the temporal relevance condition to get respective storage lifetime of event tuples for target event queries. This topic is discussed in section 3 and 4, and is optimization for memory consumption of event query evaluations.

2 Basic Framework

Events are defined as activities of interest in a system[4]. All events signify certain activities, but they have different meaning in hierarchy and semantic level. We get concepts as primitive event and complex event. Complex events basically form an event hierarchy in which complex events are generated by composing primitive or other complex events using a set of complex event operators.

In our event stream model, we assume time is a discrete, ordered domain, isomorphic to the domain of natural numbers. So we can represent timestamps as natural numbers. All events have an element called event interval that indicates its occurrence period. A time interval i consists of two timestamps $i.s$ and $i.e$, which are the end points of the interval, called the interval's start time and end time. We assume primitive events are instantaneous and complex events are durative. Thus for primitive events, this interval is a single point (i.e., identical start and end time) at which the event occurs. For complex events, the intervals contain the time intervals of all sub-events.

2.1 Complex Event Operators

We support operations over events like $A+B$, $A|B$, AB, $A;B$, A^*, $A-B$, $A||B$, A^N, A^G, A_T, which respectively represent the conjunction, disjunction, concatenation, sequence, concurrency, iteration, negation, predicate selection, aggregation and time restriction of events[5]. Due to the interval semantic, the language consist of these operators is composable other than the famous SASE[2] and SASE+[6] language. So we can construct hierarchical complex event expressions using these operators recursively. For detail semantics of the language, please refer to our work in [5].

2.2 System Architecture

We integrate ESP feature into our Service-Oriented Computing Platform InforSIB[5]. The ESP encompasses event expression algebra, a language implementing this algebra and runtime which is build for the efficient execution of event queries over vast amounts of events, Fig. 1 shows those parts schematically.

At compile time, the front-end query optimization takes care of parsing complex event queries. If with temporal restriction constraints, event expressions are first converted to alternative but easy detected form. Then expressions with outmost negation operator are rewritten into expressions with predicates injected into the denotation of operators in right places. At last we compute the temporal relevance of primitive events to the registered queries. An arbitrarily complex event query is converted to an event tree. There is one node in the tree for every temporal related and logic operator (sequence, its variants like concurrency, and logic like conjunction) in the query and

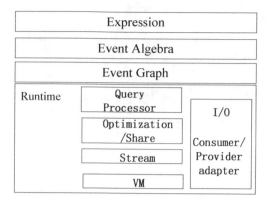

Fig. 1. System Hierarchy for Event Stream Processing

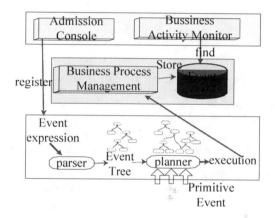

Fig. 2. Implementation Framework for Event Detection

the structure of the tree is derived from the hierarchical structure of the event operators in the query.

After we converted the complex event query expressions as evaluate-friend form,
we generate plans for these queries with event trees input shown in Fig. 2. Common
events in different event trees, which we called here shared events, are merged to
form nodes with ancestors in an event graph. Nodes in an event graph are either
primitive nodes or operator event nodes come from event trees. The inputs to operator
nodes are either complex or primitive events and their outputs are complex event. The
query plan runs 3 steps as in Fig. 3, and the sample query is semantic equal to the
SASE's example query plan in [2]. We evaluate complex event queries apparently
different in the selection step. The selection step has 3 parts to evaluate in order, first
is results from predicate injection into the negation as explained in Section 3.2, then
the temporal restriction in Section 3.1, at last we get the normal selection as in SASE.
We treat the negation and window (we call it temporal restriction here) as selection
and evaluate it as early.

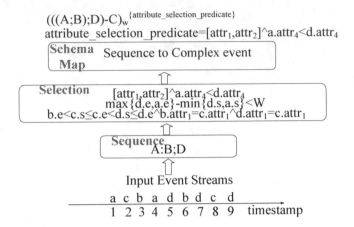

$(((A;B);D)-C)_w^{\{attribute\_selection\_predicate\}}$
attribute_selection_predicate=[$attr_1,attr_2$]$^\wedge$a.$attr_4$<d.$attr_4$

Fig. 3. Query Plan for Complex events

3 Query Conversion

We have two forms of query conversion. The first is to convert event queries with temporal restriction operators to alternative but easy detected form; this is based on the algebraic equality of event expressions under the same input. We further convert the queries with rewrite rules, which convert the queries with negation as outmost operator to forms with predicates injected into the denotation of operators in right places.

3.1 Algebraic Conversion

Every primitive event E_i(identity of i) has a schema R_i, $R_i=\{i.s, i.e, x_1,..., x_n\}$.We define the occurrence time of a tuple e in the result of expression E is the latest time-stamp in e. For event expressions A and B, we define $A\equiv B$ if A and B have the equivalent denotation under the same input stream. Trivially, '$\equiv$' is an equivalence relation. We can easily get forms of equivalence as below:

$$1. A \equiv A_T \text{ Aisprimitiveevent}$$
$$2. (A|B)_T \equiv (A_T|B)_T \equiv (A|B_T)_T \equiv A_T|B_T$$
$$3. A_T|B_{T'} \equiv (A_T|B_{T'})_{\max(T,T')}$$
$$4. (A;B)_T \equiv (A_T;B)_T \equiv (A;B_T)_T$$
$$5. (A+B)_T \equiv (A_T+B)_T \equiv (A+B_T)_T$$
$$6. (A-B)_T \equiv A_T - B \equiv A_T - B_T \equiv (A-B_T)_T$$
$$7. (A_T)_{T'} \equiv (A_{T'})_T \equiv A_{\min(T,T')}$$

Based on the equations above, we can convert event expressions into evaluate-friend forms, which mean they can be evaluated with less event instances and terminate the evaluate as early too. First, the conversion process is automatic and it treats temporal restriction and restriction-free expressions uniformly, i.e. we define $E= E_\infty$. Second, if $B\equiv B_T$, we can convert sequence $A;B$ to the equivalent form of $A;_T B$, we can see from the detection algorithm from [5] that $A;_T B$ can be detected with less memory and its detection can be terminated as early. And more, as we will explain in Section 4, $A;_T B$ has short relevance time than $A;B$, so it can be used later during the compile time to expire unnecessary reference to event tuples. We present the conversion algorithm in Table 1. Evidently the time complexity of this algorithm is linear to the size of input event expression.

Table 1. Conversion Algorithm for Temporal Restriction

Algorithm TSConvert
Input: Event expression E, Temporal Restriction T
Output: Converted target expression E', Converted target
temporal restriction T''

1.	CASE E of
2.	$[E \in P]$:
3.	$E'=E;T'=0$
4.	$[A|B]$:
5.	$<A',T_a>=$TSConvert(A,T)
6.	$<B',T_b>=$TSConvert(B,T)
7.	$E'=A'|B';T'=\max(T_a,T_b)$
8.	$[A+B]$:
9.	$<A',T_a>=$TSConvert(A,T)
10.	$<B',T_b>=$TSConvert(B,T)
11.	$E'=A'+B';T'=\infty$
12.	$[A-B]$:
13.	$<A',T_a>=$TSConvert(A,T)
14.	$<B',T_b>=$TSConvert$(B,\min(T_a,T))$
15.	$E'=A'-B';T'= T_a$
16.	$[A;B]$:
17.	$<A',T_a>=$TSConvert(A,T)
18.	$<B',T_b>=$TSConvert(B,T)
19.	IF $T_b \leq T$
20.	$E'=A';_{Tb} B';T'=\infty$
21.	ELSE
22.	$E'=A';_T B';T'=\infty$
23.	$[A_{T'}]$:
24.	$<A',T_a>=$TSConvert$(A,\min(T,T''))$
25.	IF $T_a \leq \min(T,T'')$
26.	$E'=A';T'= T_a$
27.	ELSE
28.	$E'=A'_{T'};T'= \min(T,T'')$

3.2   Rules for Rewriting

Given that the expression specified in temporal related operators can be very complex and may involve multiple levels of negation, it becomes quite hard to reason about the semantics of event expression if there exist selection constraints specified as A^N. Assume the event expression is A^N where N is a selection predicate P and the A specified form is a complex event E, thus we denote it an algebraic alternative form SELECT $\{P\}(E)$. SASE inject predicates in N into the denotation of temporal related operators, and the position of injection depends on whether the operators involve negation or not[2]. Predicate injection require that the selection predicate P to rewritten as a disjunctive normal form $P=P_1 \vee P_2 \vee ... \vee P_k$, where each P_i is a conjunction. Further we rewrite P_i as $P_{i+} \wedge P_{i-}$, with P_{i+} denoting the conjunction of those predicates that do not involve a variable referring to a negative component, and P_{i-} representing the rest.

Since there is no satisfactory semantics in the case when selection predicate P_i is not equivalent to $P_{i+}{}^{\wedge} P_{i-}{}^{[7]}$, we therefore regard such queries as invalid inputs and will reject them upon query registration. As CEDR pointed out, problems will occur during rewriting if we use the same as SASE when outmost negation operator exists[7]. We modify the rewrite rules as below:

If the top level operator in temporal related operators is not a negation, we can proceed in a way similar to SASE as Rule1.

Rule1:SELECT$\{P\}(E)$=SELECT$\{P_1 \vee P_2 \vee ... \vee P_k\}(E)$=SELECT$\{P_1\}(E) \cup$SELECT$\{P_2\}(E) \cup ... \cup$SELECT$\{P_k\}(E)$.

When negation is the outmost operator, we get Rule2.

Rule2:SELECT$\{P_+{}^{\wedge}P_-\}(E_1$-$E_2))$->SELECT$\{P_-\}(E_1)$-SELECT$\{P_+\}(E_2)$.

Rewriting in Rule2 is not bidirectional, just a uni-directional process to inject predicates into the denotation of operators in right places. The injection process is recursive, and when it gets down to the primitive event instead of a complex event expression, we inject P- into the original denotation of negation operator.

4   Determine Temporal Relevance of Event to Queries

4.1 Computing Temporal Relevance Conditions

After the algebraic conversion and rewriting, complex event expressions are now with temporal and non-temporal constraint selection conditions. In this subsection, we will discuss how to compute the lifetime of event reference relevant to a given query.

The CERA[8] treats temporal operation as selection constraint combined with join, and they propose an approach to static determine temporal relevance of event to queries, remove event tuples from the histories when know for sure that the tuples will not affect any future answers. We adopt similar approach but differ in the data model. In this paper, for all the complex event expressions to be evaluated, each input event just has one instance, the shared parts (primitive event and intermediate complex event) will be a reference to the instance and the lifetime of the reference is the same with the temporal relevance in CERA, while the instance's lifetime is determined by the times reference to it and the reference's lifetime, i.e. a reference count approach. The intermediate complex events reference to the instance by pull or by push that the instance initiates.

For a given complex event query, there is a set of temporal relevance conditions, one for each of input relation R of each shared intermediate uncompleted complex event. Temporal relevance conditions have the form as:

$$TR_{R\ in\ Q} \equiv i_1 \geq now - rt_1{}^{\wedge}...{}^{\wedge}i_m \geq now - rt_m{}^{\wedge}\ i_{m+1} > now - rt_{m+1}{}^{\wedge}...{}^{\wedge}\ i_n > now - rt_n$$

All rt_i are fixed lengths of time and can be treated as individual temporal restriction for each timestamp in the condition.

To compute the temporal relevance condition of each complex event queries, i.e., for a given expression E, compute the relevance time $rt_{R\ in\ E}(i)$ of each timestamp i of an input relation in E. And, for each relevance time, we can compute the temporal distances $td(i,j)$, which is the least upper bound the temporal distance from i to j, then we can choose the longest of all temporal distances $td(i,j)$ from i to other timestamp j of E.

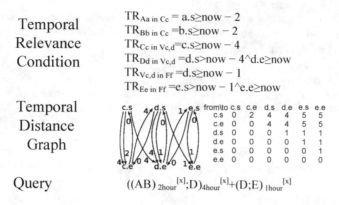

Temporal Relevance Condition

$TR_{Aa \text{ in } Cc} = a.s \geq now - 2$

$TR_{Bb \text{ in } Cc} = b.s \geq now - 2$

$TR_{Cc \text{ in } Vc,d} = c.s \geq now - 4$

$TR_{Dd \text{ in } Vc,d} = d.s > now - 4 \wedge d.e \geq now$

$TR_{Vc,d \text{ in } Ff} = d.s \geq now - 1$

$TR_{Ee \text{ in } Ff} = e.s > now - 1 \wedge e.e \geq now$

Temporal Distance Graph

from\to	c.s	c.e	d.s	d.e	e.s	e.e
c.s	0	2	4	4	5	5
c.e	0	0	4	4	5	5
d.s	0	0	0	1	1	1
d.e	0	0	0	0	1	1
e.s	0	0	0	0	0	1
e.e	0	0	0	0	0	0

Query

$((AB)_{2hour}{}^{[x]};D)_{4hour}{}^{[x]}+(D;E)_{1hour}{}^{[x]}$

Fig. 4. Computing Temporal Relevance Condition for Complex Event Query

Since all temporal condition in selection are equivalent to a conjunction of comparisons of the form $j-i<t$ or $j-i\leq t$, computing all temporal distances in an expression E is just turn out be an "all pairs shortest paths" problem in a directed, weighted graph. In this temporal distance graph (TDG) CERA called, the nodes are all the timestamps (the date model, not the value) occurring in E. Every selection condition from temporal restriction $j-i\leq t$ will generate a directed edge from node i to j with weight t. For each pair of one interval $i.s$ and $i.e$, we generate an edge with weight 0 from $i.e$ to $i.s$. Then the temporal distance $td(i,j)$ corresponds to the shortest path from i to j in the graph. All the shortest paths between all timestamps in E can be computed using standard algorithm, at last we get relevance time $rt(i)$ for i as the maximum entry in the row where i stands.

The expression represented in our event algebra $((AB)_{2hour}{}^{[x]};D)_{4hour}{}^{[x]}+(D;E)_{1hour}{}^{[x]}$ is equivalent to example query in CERA. We name the intermediate sub-event $(AB)_{2hour}{}^{[x]}$ as C, and $(C;D)_{4hour}{}^{[x]}$ as V and the whole expression as F. The temporal relevance conditions are computed as in Fig. 4.

4.2 Event Tree

After the computing of temporal relevance conditions, we get event trees for each event query.

The leaf nodes in the event tree are primitive event nodes. A primitive event node exists for each primitive event type and stores references to the instances of that

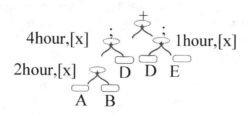

4hour,[x]   1hour,[x]

2hour,[x]   D   D E

A B

Fig. 5. Event Tree

primitive event type. Operator nodes execute the event operators on their inputs, and they are the non-leaf nodes. The event tree is executed in a bottom-up manner. The event tree parsed from event expression in section 4.1 is shown in Fig. 5.

4.3 Memory Management for Event Tree

As a space optimization, we do not materialize each copy of a base event inside the operator independently. Here, only a copy of the event is maintained in a Primitive Event Cache and intermediate events consist of pointers to events in this cache. If multiple queries have common parts in their event trees, the respective intermediate events will be materialized to be shared among these queries. Every materialized intermediate event has information about the relevance time of events to it (square beside the tree node in Fig. 6). Every logic and temporal related complex event operator has its tailored strategy to combine this information into his memory management framework, thus make ESP runtime know how to handle these event tuples' lifetime and abandon them at exact time.

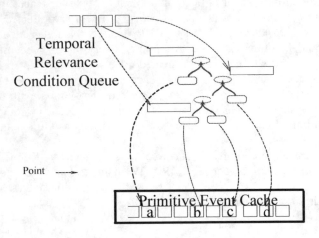

Fig. 6. Memory Management for Event Tree

5 Experiments

In this section, we will present the experimental results on the efficiency of our query optimization approach. Our experiments consist of two parts to evaluate the query performance, the throughput and processing latency of query execution. All experiments were conducted on a 2.8 GHz PentiumIV PC with 1GB memory, running Windows XP Professional SP2 with JDK 1.6.0_02. All experiments are averaged over 3 runs.

Events out of order or irrelevant to the registered queries are called irrelevant event or noise event. The two variables in these tests are: the total number of events send and the irrelevance ratio. The two target ESPs for tests are the simple ESP framework in [5] and the one with query optimization presented in this paper. When needed, the JVM tuning will be set to "-Xms512m –Xmx512m -Xns128m -XX:+UseParNewGC", to trigger the Garbage Collection.

5.1 Throughput

Two factors are measured: total time to complete test and the memory usage of the ESP framework. The total time to complete test is the sum of time needed to send all the events and the time needed to do the callbacks when complex events are detected.

The tested event queries are 4 queries in section 5.2 with the varying send times of irrelevant input events D of different value of x. That is, we send M times (A(1), B(1), N times of D(1), D(2), E(1)) with the desired temporal restrictions and N is the varying times of sending D.

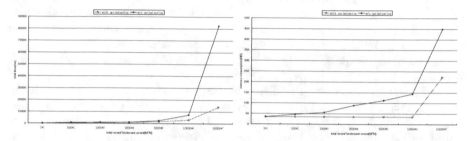

Fig. 7. Throughout of Event Query Evaluation: Total Processing Time and Memory Consumption

We can see the improved performance in Fig. 7: above 2 times faster and saving memory consumption from 23% up to 75%.

5.2 Latency

Latency for ESP processing is the total time to completely detect a complex event, from sending the first event to the ESP until a callback is received when the complex event fires a match.

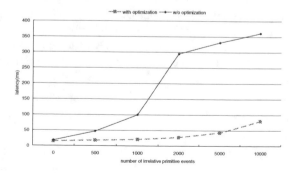

Fig. 8. Latency of Event Query Evaluation

The tested event query is $((AB)_{2hour}{}^{[x]}; D)_{4hour}{}^{[x]} + (D; E)_{1hour}{}^{[x]}$ with the varying send times of irrelevant input events D of different value of x. That is, we send A(1), B(1), N times of D(1), D(2), E(1) with the desired temporal restrictions and N is the varying times of sending D. The query optimization can obtain roughly linearly curve of latency lower than the optimization-free one as in Fig.8.

6 Related Work

We study optimization for temporal restriction queries in ESP. Temporal restriction query as a key component and multi-query optimization (MQO) in ESP are interesting and emerging research areas for ESP performance and scalability.

The conventional approaches for stream query processing are to use selection-join-aggregation queries[9, 10], but they are inconvenient and the unintuitive queries are hard to write and understand. It is not suitable to adopt these conventional approaches in ESP. Recent studies have started to address efficient evaluation of ESP queries [2, 3, 6], and tailored query optimization [8, 11, 12] are presented to obtain high-performance query for ESP.

SASE[2] and its successor[6] are NFA-based ESP engine that uses a stack implementation to filter and correlate events. They present optimizations for handling large windows and reducing intermediate results[2] and exploit aggressive sharing in storage of all possible complex event matches[6]. As for the expressibility of query language, they have to be further extended in order to support full compositionality in these languages.

Cayuga demonstrates the use of their NFA-based ESP for Web feeds and stock prices[3]. It can run multiple queries concurrently and events sequences are implemented by NEXT-FOLD constructs. Cayuga provides an implementation and a set of well documented patterns used for ESP implementations. Those patterns are mainly concerned with efficient execution of queries, indexing, data management, and garbage collection. But it doesn't output the complete matches.

7 Conclusions

In this paper, we propose a query planed approach to evaluate complex event queries in ESP. The query optimization approaches in ESP framework achieve performance gains. As for further performance improvement, we will leverage more potentialities for optimization, including runtime query optimization by exploiting event relevance constraints as in [12].

Acknowledgement

The authors gratefully acknowledge financial supports from Project 863 under granted number 2006AA01Z451,2007AA01Z474, the Ministry & Commission-Level Research Foundation of China under granted number [2006]634.

References

[1] Gold-Bernstein, B., et al.: SOA Trends: Intersection of SOA, EDA, BPM, and BI (2007), http://www.ebizq.net/filelib/8692.html

[2] Wu, E., et al.: High-Performance Complex Event Processing over Streams. In: Proc. SIGMOD, pp. 407–418 (2006)

[3] Demers, A., et al.: Cayuga: A General Purpose Event Monitoring System. In: Proc. Biennial Conference on Innovative Data Systems Research, CIDR (2007)

[4] Luckham, D.C.: The Power of Events: An Introduction to Complex Event Processing in Distributed Enterprise Systems, p. 376. Addison-Wesley, Reading (2001)

[5] Jia-hong, L., Quan-yuan, W.: An Event-Driven Service-Oriented Computing Platform. Chinese Journal of Computers 31(4), 588–599 (2008) (in Chinese)

[6] Agrawal, J., et al.: Efficient Pattern Matching over Event Streams. In: Proc. ACM SIGMOD (2008)

[7] Barga, R.S., et al.: Consistent Streaming Through Time, Technical Report, Microsoft Research (2006)

[8] Bry, F.o., Michael, E.: On static determination of temporal relevance for incremental evaluation of complex event queries. In: Proc. International Conference on Distributed Event-Based Systems(DEBS), pp. 289–300. ACM, New York (2008)

[9] Arasu, A., Babu, S., Widom, J.: CQL: A language for continuous queries over streams and relations. In: Lausen, G., Suciu, D. (eds.) DBPL 2003, vol. 2921, pp. 1–19. Springer, Heidelberg (2004)

[10] Chandrasekaran, S., et al.: TelegraphCQ: Continuous Dataflow Processing for an Uncertain World. In: Proc. Biennial Conference on Innovative Data Systems Researches, CIDR (2003)

[11] Wei, M., et al.: ReCEPtor: Sensing Complex Events in Data Streams for Service-Oriented Architectures, HPL-2007-176, Digital Printing and Imaging Laboratory,HP Laboratories,Palo Alto (2007)

[12] Ding, L., et al.: Runtime Semantic Query Optimization for Event Stream Processing. In: Proc. ICDE, pp. 676–685. IEEE Computer Society, Los Alamitos (2008)

Video Object Segmentation Based on Disparity

Ouyang Xingming and Wei Wei

Computer College of Science & Technology, Huazhong University of Science & Technology,
Wuhan 430074, China
ouyangxingming@163.com

Abstract. Disparity information enables a multi-layered representation of a video frame. But how to extract object attracting the users' interest is still to be considered. Based on the definition and estimation of image disparity, we propose an approach to segment video objects for video conference. Chrominance, luminance, position and other information are introduced to compute the saliency map, and the edge information is used to obtain the disparity map. Combining with the disparity estimation results, the layer representing the most interested object that contains most of conspicuity values can be easily segmented. The experimental results demonstrate this algorithm is capable of segmenting video object from sequences quite effectively.

1 Introduction

Although most video archives mainly consist of 2-D video sequences, the use of 3-D video, obtained by stereoscopic or multiview camera systems, has recently increased since it provides more efficient visual representation and enhances multimedia communication. 3-D video enables users to handle and manipulate video objects more efficiently by exploiting, for example, depth information provided by stereo-image analysis. Furthermore, the problem of content-based segmentation is addressed more precisely since video objects are usually composed of regions belonging to the same depth plane [1]. Various applications, such as video surveillance, image/video indexing and retrieval, or editing of video content, can gain from such 3-D representation. For this reason, 3-D data acquisition and display systems have attracted great interests recently and consequently archives of 3-D video information are expected to rapidly increase in the forthcoming years.

Depth information would enable a multi-layered representation of a video frame. Each layer would contain image regions that are at a specific distance from the video camera. The depth information of a scene is contained in a so-called depth map. However, depth information is difficult to estimate. Compared with depth estimation, disparity is more easily obtained, and it is mainly used in video transmission and 3D reconstruction. In this paper, visual attention concept is considered. The saliency map represents human concentration grade is calculated with multiple features like chrominance, luminance, position and etc. It is integrated with disparity information to extract most interested object from video sequences. The experimental results demonstrate the capability of our algorithm.

L. Chen et al. (Eds.): APWeb and WAIM 2009, LNCS 5731, pp. 36–44, 2009.
© Springer-Verlag Berlin Heidelberg 2009

The rest of this paper is organized as following. Section 2 gives a brief introduction of disparity definition and estimation approach. Section 3 proposes an approach to extract video object based on disparity. Section 4 demonstrates the effectiveness of the proposed approach. Section 5 is the conclusion.

2 Disparity Definition and Estimation

2.1 Disparity Definition

Disparity is defined as the relative displacement between the left and the right image points belonging to the same object point, as defined in Fig. 1. Without first needing camera calibration, it will find more applications in generic video communications. We will investigate how disparity information can be incorporated into existing video segmentation techniques, which would pave the way to developing novel techniques that will provide a significant improvement over existing methods.

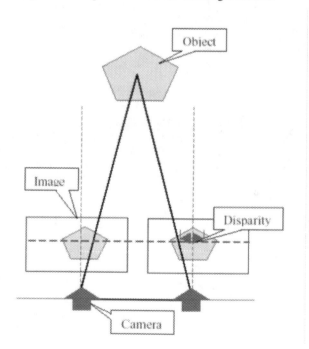

Fig. 1. Definition of Disparity

2.2 Disparity Estimation

It is well known that disparity estimation is still a hard problem, and much progress has been made over the past 25 years or so. It remains a difficult vision problem for the following reasons: noise, textureless regions, depth discontinuities and occlusions. The algorithm proposed in [2] is used to obtain the disparity map. Compared with other disparity estimation algorithm, it uses edge propagation and interpolation method, without using block matching, the boundaries of objects have no blocky results and all the pixels in the same region have similar disparities.

Disparity information enables a multi-layered representation of a video frame. But how to extract object attracting the users' interest is still to be considered. Visual attention is a neurobiological concept having the ability to concentrate the mental power upon an object on close observation. Different applications may have different attention model definitions. Moving object segmentation approaches assume that objects of interest have distinct motion with background. Photographers always think the most important objects should to be located in the center of the image. Video surveillance system takes human as interesting objects.

Multiple features in the following formula are used to generate the saliency map (SM), which indicates the saliency at every location in the visual field by a scalar quantity.

$$SM = f(depth, color, contour, texture, size,$$
$$location, motion....)$$
(1)

Layer L_i represents layer i, when L_i contains the largest conspicuity value S_{Li}, L_i will be 1 and extracted.

$$L_i = \begin{cases} 1, & \text{if } S_{L_i} = Max(S_{L_i}) \\ 0, & \text{otherwise} \end{cases}$$
(2)

$$S_{Li} = \frac{\sum SM_{Li}}{N(L_i)}$$
(3)

Here SM_{Li} denotes the conspicuity value of every layer L_i and $N(L_i)$ is the total pixels number of every layer L_i.

3   Video Object Segmentation Based on Disparity

3.1   Existing Methods and Their Restrictions

The motivation of segmentation head-and-shoulder type video sequences stems from the popular presence of head-and-shoulder type video signal in real-time services such as videophone and web chatting is to extract the person in front of the camera. Though segmentation of head-and-shoulder sequence is very useful, among the algorithms we reviewed so far, most algorithms are based on face detection and motion information, whose results are not good enough. Moving object segmentation uses motion information as rule to group regions. It assumes that physical objects are often characterized by a coherent motion, which is different from that of the background. As a result, these kinds of algorithms primarily take advantage of motion information, and are not able to handle interesting objects without motion. If the shoulder of human is stationary, it will be lost throughout the sequence. To illustrate this problem, an example is shown in Fig. 2. In Grandmother sequence, only the head of the woman has motion. So if use the motion as the only cue to extract objects, the body below the head will be lost.

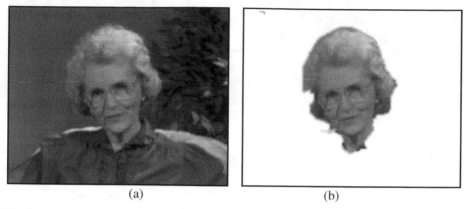

Fig. 2. Segmentation results of the Grandmother sequence: (a) original frame, (b) segmentation results

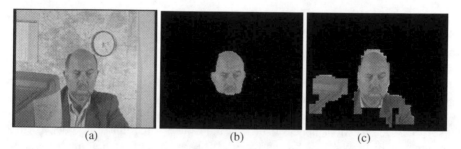

Fig. 3. Claude stereoscopic image pair: (a) Original image, (b) segmentation result of [5], and (c) segmentation result of [3]

Using color information is another method to detect human. These kinds of algorithms can only detect skin color pixels, like the face and hand parts, which have been discussed in [3-5]. Fig. 3 (b) and (c) show the segmentation results of algorithm in [3] and [5]. But in common sense, human's head and shoulder can not be separated, they should be the parts of one object. So in some applications, like video surveillance, these methods will fail.

3.2 Object Segmentation in Video Conference

Disparity information enables a multi-layered representation of a video frame. Since video objects are usually located on the same depth plane, so depth segmentation provides meaningful frame content representation. The advantage of using disparity is robust to motion fluctuation, even the object stays still for arbitrarily long period of time or when its different parts exhibit different motion characteristics.

In a video conference sequence, the main object of interest is always the person in front of the camera. In this case, the general position of the person and the skin color become very important information in extracting the object (the person's head and shoulder). Face saliency map (FSM) proposed in [5], which considered chrominance, luminance and position information, will indicate the human position.

Assume (x, y) represents the spacial position of a pixel in the current image. The corresponding luminance and chrominance components of the pixel are denoted by $Y(x, y)$, $Cb(x, y)$, and $Cr(x, y)$, respectively. The FSM can be defined as

$$FSM(x, y) = P_1(x, y) \cdot P_2(x, y) \cdot P_3(x, y) \tag{4}$$

where P_1, P_2, and P_3 denote the "conspicuity maps" corresponding to the chrominance, position, and luminance components, respectively.

Chrominance Conspicuity Map (CCM) P_1: It is known that face region generally exhibits the skin-color feature. Therefore, using the skin-color information, the facial saliency map can be easily constructed to locate the potential face areas. The P_1 is defined as

$$P_1(x, y) = \exp\{-(\omega_{Cr}(x, y)\frac{Cr'(x, y)^2}{2\Delta_{Cr}^2} + \omega_{Cb}\frac{Cb'(x, y)^2}{2\Delta_{Cb}^2})\} \tag{5}$$

$$Cr'(x, y) = (Cr(x, y) - \mu_{Cr})\cos(\theta) + (Cb(x, y) - \mu_{Cb})\sin(\theta) \tag{6}$$

$$Cb'(x, y) = (Cb(x, y) - \mu_{Cb})\cos(\theta) + (Cr(x, y) - \mu_{Cr})\sin(\theta) \tag{7}$$

where $\mu_{Cr} = 153$, $\mu_{Cb} = 102$, $\Delta_{Cr} = 20$, $\Delta_{Cb} = 25$, $\theta = \frac{\pi}{4}$ and $\omega_v(x, y)$ ($v = Cr$ or Cb) is a weight coefficient.

Position Conspicuity Map (PCM) P_2: In typical head-and-shoulder video sequences, most of the face locations appear at or near the center of the image in order to attract user attention distinctly. Few human faces are captured and placed at the boundary of the image, especially the bottom of the image. Hence, it is reasonable to assume that the probability of the face pixels existing at the center of the image will be larger than other locations. Let H and W denote the height and width of the image, respectively. Based on this characteristic, the Position Conspicuity Map P_2 is defined as

$$P_2(x, y) = \exp\{-\frac{(x - H/2)^2}{0.8 \cdot (H/2)^2} - \frac{(y - W/2)^2}{2 \cdot (W/3)^2}\} \tag{8}$$

Luminance and Structure Conspicuity Map (LSCM) P_3: From the histogram of Y component for the facial test data, the region of [128-50, 128+50] tends to contain most of conspicuity values for the facial skin area. The darker the intensity value of a pixel, the less possible will be a skin-tone color. Similar result can also be found for the very bright pixels. With these observations, LSCM P_3 is defined as

$$P_3(x, y) = s \cdot \exp\{-\frac{(\gamma(x, y) \cdot Y'(x, y) - \mu_L)^2}{2 \cdot \Delta_L^2}\} \tag{9}$$

where $\mu_L = 128$, $\Delta_L = 50$, $\gamma(x, y)$ denotes the luminance compensation coefficients. Here s denotes the structural coefficient, which is employed to characterize the luminance variation in face region.

(a)   (b)   (c)

Fig. 4. Claude stereoscopic image pair: (a) Original image, (b) SM, and (c) disparity map

According to (4), the final *FSM* can be easily obtained by employing the three conspicuity maps (5), (8), and (9).

With the observation of above *FSM* map, it can be found that in the face region, the pixel's *FSM* value is larger and pixels with large value are more centralized. Combing with the disparity estimation results, the layer representing human tends to contain most of conspicuity values for the facial skin area.

$$SM_{Li} = \sum FSM(x,y) \times w_f(x,y), \quad (x,y) \in L_i \tag{10}$$

where $FSM(x, y)$ is the face saliency value of pixel (x, y), w_f is the a weight coefficient, which is employed to characterize the density of bright pixels in neighboring region of pixel (x, y). The larger density value and the larger *FSM* value of pixels, the larger conspicuity value SM_{Li} will be. In order to avoid the influence of discrete bright pixel, the weight coefficient w_f is defined as

$$w_f(x,y) = \frac{\sum_{k,l=-r}^{r} B(FSM(x-k,y-l),T)}{r^2} \tag{11}$$

$$B(FSM,T) = \begin{cases} 1, & FSM \geq T \\ 0, & \text{otherwise} \end{cases} \tag{12}$$

$$T = \tau \cdot F_{max} \tag{13}$$

$$F_{max} = \max(FSM) \tag{14}$$

where τ is a threshold. According to the performance of a lot of experiments, the value of 0.7 is recommended, which can provide better constraint result for the candidate face pixels selection. Only the pixels with *FSM* value larger than the threshold will be considered, the pixels with small *FSM* will not influence the conspicuity value of every layer. When we calculate w_f of pixel (x, y), only $r \times r$ neighborhood of pixel (x, y) are considered. From (11), we can see that only those pixels with larger *FSM* values and higher weight coefficient will be taken into account, which means that the influence yielded by the discrete pixels with large *FSM* values and consecutive pixels with small *FSM* values will be reduced significantly.

4 Experimental Results and Analyses

Claude stereoscopic image pair is a typical video conference scene. Figure 4 (b) is the *SM* of stereoscopic image Claude, and the layer representing human is depicted in Fig. 5(c).

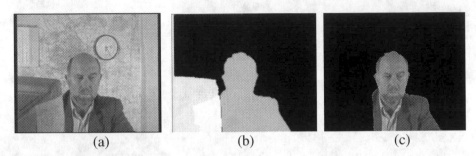

(a)   (b)   (c)

Fig. 5. Claude stereoscopic image pair: (a) Original image, (b) disparity map, and (c) layer representing human

Disparity can not only be used to segment video conference image, but also be utilized to segment other video images. To illustrate the effectiveness of our algorithm, more experimental results have been given in Figure 6~ 8.

(a)   (b)   (c)

Fig. 6. Image pair of Flower Garden: (a) original image, (b) disparity map, (c) center object

Figure 6 and 7 show two examples of center object segmentation. When camera moves to capture object of interest, the most important objects should usually be located in the center of the image. So segmentation of this kind of sequence, the position is the most important feature. The saliency map (SM) can be defined as

$$SM = f(position) \tag{15}$$

The object in the middle of scene can be extracted, as illustrated in Fig. 6 (c) and Fig 7 (c).

Figure 8 is an example of nearest object segmentation. If we pay more attention to nearest objects, the *SM* should have higher value in near object. Position feature is the most important cue. *SM* is defined as following:

$$SM = f(depth) \tag{16}$$

Fig. 7. Stereoscopic image pair of Piano: (a) original image, (b) disparity map, (c) center object

Fig. 8. Synthesized stereoscopic image pair: (a) Original image, (b) disparity map, (c) nearest object

Fig. 9. Claude stereoscopic image pair: (a) Original image, (b) nearest object

According this kind of application, the nearest object will be extract. Figure 8 is a synthesized stereoscopic image pair. There are several objects in the scene, and the nearest object is shown in Figure 8(c). Figure 9 depicts the result of nearest object extraction. Differed from the result in Figure 5, the extracted object is the computer not the human.

All the above experimental results have demonstrated that our method is capable of segmenting video object quite effectively.

5   Conclusion and Future Work

In this paper, we propose an approach for video object segmentation based on disparity information and visual attention. Disparity can be an important feature, which provides layer information to segment video object from sequence. According to different applications, experimental results proved its efficiency. However, the disparity estimation proposed in the reference [2] has two limitations. One is that for a region without matching feature points (usually this kind of region has no texture information) the algorithm cannot find the accurate disparity value inside such region. This also exists in other stereo matching algorithms. Another limitation is that the interpolation method is based on the color information. When foreground and background objects have similar color or objects have gradually changing color, the algorithm will not work as well. Our algorithm based on this disparity estimation algorithm is time consuming. To achieve a real-time computation target, many processes in the algorithm should be optimized in future works.

References

1. Garrido, L., Marques, F., Pardas, M., Salembier, P., Vilaplana, V.: A hierarchical technique for image sequence analysis. In: Proc. Workshop Image Analysis for Multimedia Interactive Services (WIAMIS), Louvain-la-Neuve, Belgium, June 1997, pp. 13–20 (1997)
2. Wei, W., Ngan, K.N.: Disparity estimation with edge-based matching and interpolation. In: IEEE International Symposium on Intelligent Signal Processing and Communications Systems, Hong Kong SAR, China, December 2005, pp. 153–156 (2005)
3. Chai, D., Ngan, K.N.: Face segmentation using skin-color map in videophone applications. IEEE Trans. On Circuits and System for Video Technology 9, 551–564 (1999)
4. Habili, N., Lin, C.C., Moini, A.: Segmentation of the Face and Hands in Sign Language Video Sequences Using Color and Motion Cues. IEEE Trans. On Circuits and System for Video Technology 8, 1086–1097 (2004)
5. Li, H., Ngan, K.N.: Face Segmentation in Head-and-Shoulder Video Sequences Based on Facial Saliency Map. In: IEEE International Symposium on Circuit and Systems (2006)

A Database Approach for Accelerate Video Data Access

Wenjing Zhou[1,2], Xiangwei Xie[1,2], Hui Li[1,2], Xiao Zhang[1,2], and Shan Wang[1,2]

[1] Key Laboratory of Data Engineering and Knowledge Engineering
(Renmin University of China), MOE, Beijing 100872, P. R. China
[2] School of Information, Renmin University of China, Beijing 100872, P. R. China
{Zhouwj0507,Xiexw,HLi,Zhangxiao,SWang}@ruc.edu.cn

Abstract. In this paper, we studied the efficiency and break-event point of storing video objects into DBMS and proved that storing "small" video objects into database is a suitable solution. To the video objects that stored in database as BLOB data type, we devised a database based time-oriented approach to speed up the video content access. Our experiments showed that, because of we extracted some system-aware metadata and stored into database transparently, the read performance was become practicable.

Keywords: Video Data Storage, Video Data Access, Binary Large Object.

1 Introduction

The volume of video data increased rapidly since last decade [1]. Video becomes popular in our everyday life for professional and consumer applications, e.g. surveillance, education, entertainment. Such needs require the video data management system should provide a mechanism to store and access the data in an effective way.

The existing video storage strategies are usually divided into two categories: file system based video data store and database LOB (Large Object) data type based video data storage. Most of video surveillance, video-on-demand and online video sharing system are adopting the former. In this approach, the UNC path of video files or other location description information are stored into database. While processing a video data access request, the applications should get the data file location information from database before performing the corresponding data access operation. This mechanism simplified the video data management into video data file location information management. However, there are remaining some disadvantages in adopting the file based video storage strategy. Firstly, because of the content files are outside database, it is hard to coordinate them with its location information and other metadata in consistency. Secondly, putting a large number of video thumbnails and content files in a directory will result in inefficiency for responding data request. In fact, it has been a bottleneck in YouTube until it was bought by Google and employed the GFS[2] as the underlying infrastructure [3]. Thirdly, developing the GFS-like solution start from scratch, or deploying the large-scale video application underlying the similar open source system, such as Hadoop [4], is still a difficult task.

Another video storage solution was storing video data in the BLOB field of the database. The engineering experiences have showed that, through utilizing its built-in

L. Chen et al. (Eds.): APWeb and WAIM 2009, LNCS 5731, pp. 45–57, 2009.

data management functionality and the well provided development interface in DBMS, it is easy to build a large-scale application. Unfortunately, the folklore tells us that DBMS was only efficiently in handle small objects. For the video related web application, e.g. YouTube, the largest video clips sharing web site, most of the uploaded clips are small and last 30 seconds to 2.5 minutes [5], its size is around 25 KB per second. In some others video applications, the video objects are also small, such as content based video retrieval system, their video files are segmented into shots, each shot is often less than 30 seconds. For these small video clips related applications, storing the small objects into database with a high confidence become a valid approach. But what means the "*small*" and where is the break-event point for accessing an object as BLOB in database will be cheaper than accessing an object as a file, is remain unanswered and necessary for further study.

For both storage strategies, supporting the random accessibility of the video content based on the specific time is important. It not only decreased the respond time of the request, but also saved the bandwidth for avoiding the unnecessary data buffering. To the applications involved huge volumes of data and traffics, such as video-on-demand and online video sharing, efficient use of the bandwidths has a good significance [6].

In this paper, our works were focused on database based video data storage and video content access approach. Our contributions were summarized as follows:

We investigate the break-event point in DBMS for storing the video object, we found that stored video objects into database could be a preferred solution in some circumstance, and hence is necessary to develop database based video content access approach. By evaluating long-term read/write performance, we discovered that different DBMS will vary greatly in break-event point. We also explored some factors that usually have important effects on accessing relational data, the results revealed that they hold few affection to the throughput of video objects.

We devised a time based video content access approach for the video objects in BLOB column of database. It provides a transparent way to wrapper the required content to the end user instead of sending the whole video file from beginning. Our experiments showed that, because of extracted some system-aware metadata and stored them into database in an efficient way, the performance becomes practicable.

The remainder of this paper is organized as follows. Section 2 studies the break-event point of database based video storage. Section 3 presents our database based video content access approach, followed by the reports of experiments in Section 4. Section 5 lists the related work and section 6 concludes the paper.

2   Methodology of Video Storage Evaluation

In this section, we focus our studies primarily on the methodology of performance issues and its break-event point of simple read/write operations in file system and database based storage. This study is conducted by a series of comparisons between NTFS and two commercial DBMS products in client/server style video application. The data access mode of this application is quite common and highly similar to the web based video sharing, video-on-demand, content based video retrieval, etc.

2.1 Storage Age and Safe Write

Unlike the common performance evaluation, we conduct the experiment with concern the long-run impact by introducing the concept of *storage age* [7]. It is a relative measurement of time and defined as the ratio of bytes in objects that once existed in the system to the number of bytes in current storage. It should be noticed that this definition holds the assumption where free space is relatively constant during the long-run of applications. To ensure the file based storage and database based storage is compared in the same semantics, we make all the update in the file based storage follow a way named *safe writes*, which is similar to the update of BLOB data and ensure an old file is robustly replaced. The update procedure is illustrated as follows: a new version of the object file is written to the disk with a temporary name at first, and then, it's renamed to the old object file name, finally, we delete the old file to complete this update. From the above illustration of these two concepts, we can know the storage age could be simply interpreted as the "safe write per object".

2.2 Methodology of Storage Evaluation

Our storage performance evaluation is conducted between NTFS and two commercial DBMS products on Windows platform. We chose measure the performance in a client/server style application within a single machine, i.e. the client application which issued the read and write requests, the video objects and their metadata are located in the same computer. This configuration made our experiments become simple and have no need to take account into the factor of network. For the simplicity but without loss the generality, we only pay attention to the typical read and write primitive operation in applications. Because of the update operation could be figured as compositions of delete and insert, the three primitive operations have been taken into consideration: query (select), insert, and delete.

In the database storage, the video object is stored into a BLOB column and its metadata are stored in the same table. While in the file system based storage, we stored object's UNC path and other metadata in database table, the raw data of video objects are stored as file in a directory. It should be mentioned that, in the database storage, DBMS's database files is located in an otherwise empty NTFS volume, while in the file based storage, only the raw video objects are placed in that empty disk volume.

During the performance evaluation, we only measured constant size video objects rather than objects with certain distributions in size. Intuitively, the video objects were impossible to hold a constant size in a real application scenario, and the distribution of size would play an important role to affect the performance. However, the works in [7] have showed the intuition was wrong and the object size has little impact on the performance, so we choose the constant object sizes as the basis assumption for the simplicity reason.

Under the two DBMSs, we set them in *bulk logged* mode. It means the BLOBs writing are flushed to disk at commit. This setting will make the DBMSs avoid the log write in handle the writing operations. We also set the two DBMSs under the *out-of-row* storage mode for the BLOB data, i.e., only the locator descriptor information is stored in the BLOB column, the actual data are stored in the table space out of the table, and hence the BLOB will not decluster the corresponding metadata.

To the settings of buffer, except the cases of exploit the performance impact of the *database buffer size* and *shared pool size*, we used all the default settings of the two DBMSs and both of them are 64KB, because of they are pre-fetch 8 pages (8KB per page) one time by default. For the sake of fairness, we make the read/write buffer size of NTFS storage are equal to the two DBMSs, i.e. 64KB.

We named the two commercial DBMS products as System A and System B, and then the video objects storage strategies involved in our experiments will be marked as follows:

A-LOB: Storing video objects into BLOB column in system A;
A-NTFS: Storing video objects on NTFS file system, but their metadata are in the table of system A;
B-LOB: Storing video objects into BLOB column in system B;
B-NTFS: Storing video objects on NTFS file system, but their metadata are in the table of system B.

The flow of our performance evaluation is listed below:

Step 1: Initialize the DBMS with proper configuration and connection;
Step 2: Fill the disk volume in the ratio of *50%* with constant sized video objects;
Step 3: Perform the primitive operation of *insert*, *select* and *delete* in iteration until it repeated enough times;
Step 4: Calculate the value of *storage age* and renew it while an interval of *0.5* is achieved;
Step 5: If the value of storage age is less than *10*, then go to the step 3, otherwise, the evaluation is end.

Because of the goal of our performance evaluation is studying the break-event point for database based small video objects storage that suit for the applications like online video sharing, content based retrieval and so on. The involved video object's size in our experiments are 64KB, 128KB, 256KB, 512KB and 1MB. Partial of the result and analysis are listed in section 4.1, they indicated that storing video objects into database could be a preferred solution in some circumstance, and hence is necessary to develop database based video content access approach in a system-aware way to speed up the throughput.

3   Time Based Video Content Access Approach

Section 2 and its experiments verified the efficiency and validity of storing small video objects into database. However, most DBMSs only support the access to some metadata of video file, they didn't provide the capacity of time based video content access or other types of raw data accessibility. Therefore, it is necessary to develop database based video content access approach. In our work, we implemented the functionality of accessing video data to three mainstream formats in database, i.e. AVI, MPEG-1and MPEG-4.

Generally speaking, in order to support time-based video access in database, pre-processing is needed to parse the video's format, extract and compose the temporal-index information. Extracted temporal-index information is stored into database. When a time-based video access request arrived, these extracted information were be used to locate and speed up the access. We will detail the approach in the next two subsections.

3.1 Database Based Time-Oriented AVI/MPEG-1 Content Access

The key to support time based access for AVI data is build temporal information index. The file format structure of AVI is presented on the left part of Figure 1. Sample is the lowest granularity unit for AVI content storage. The index block is an optional component. While an AVI file is loading into database as BLOB data, we will analyze the "File header" at first, which include extracting the duration time, location and length of raw video data block, then scan the "index block" to build temporal information if this component is existed, otherwise, we will scan the whole AVI file to record the position of each key sample so as to construct the temporal information index to locate the content.

File header	Sample	Sample	Sample	......	Index block		File ID	Sample ID	Is Key Frame	Sample Time	Sample Offset

Fig. 1. File format structure of AVI video and the formed temporal indexing structure

The index structure of AVI is showed on the right of Figure 1, it was stored in database as a table. "Sample ID" refers to the ID of video frames. "Is Key Frame" suggests whether the frame is a key frame. "Sample Time" tells the time at which this frame starts to be played and "Sample Offset" indicates the frame offset in video file.

While a service request that acquires partial content of AVI video is arrived, it will be handled as follows:

Step 1: Parse the request and get the ID of requested video file, the start time and termination time;

Step 2: Retrieve the time-based indexing information from database based on above three data items;

Step 3: Locate the position in LOB column according to the existed indexing information and read the corresponding raw video data;

Step 4: Rebuild the segmented content to a new file by combing file header and the wrapped raw data block based on the AVI file format standard, and then deliver the segmented clips instead of the whole file to complete this service request.

The techniques for preprocessing MPEG-1 video file, extracting temporal index information and implementing time based random-access to MPEG-1 video data in database is similar to AVI, so we didn't detail it in this paper. The biggest difference between them was there isn't specific file header in MPEG-1 video files. So we have to scan the whole file to extract the indexing information like in AVI file while there is no "index block" in the file. The time-based index structure of MPEG-1 is shown in Figure 2.

FileID	Packet ID	Packet Offset	Packet Length	Time Offset

Fig. 2. Time-based index structure of MPEG-1 file

Packet is the lowest granularity unit for MPEG-1 content storage. Figure 2 illustrated that, "Packet ID" is the ID of packets in a MPEG-1 file. "Packet Offset" indicates the starting position of each packet and "Packet Len" is the duration time of each packet in video file. "Time Offset" field denotes the start time of current packet.

3.2 Database Based Time-Oriented MPEG-4 Content Access

The file structure of MPEG-4 video files is shown in Figure 3. There are large differences in the content organization among MPEG-4, AVI and MPEG-1 files. MPEG-4 file has a complex compressed metadata structure which placed at either the front or the tail of the file. Through parsing the metadata and taking a series of complex computation, the time-based index information for key sample will be gained. Due to the complexity of compressed metadata structure, splitting MPEG-4 video file in the style that just based on the temporal index information could not ensure the segmented clips is conform with the file structure standard, it will result in the clip cannot be decode and played correctly. Therefore, in the time based MPEG-4 video content access, rebuilding new file header and metadata based on the original metadata that matched the raw content is required. In our database based approach, we implemented this functionality transparently for users.

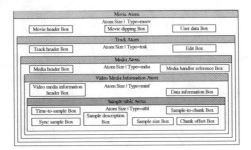

Fig. 3. File format structure of MPEG-4 file

From above illustrations and file format structure showed in Figure 3, we know that the preprocess procedure for loading MPEG-4 video file into database is much more complex than AVI and MPEG-1 video. Due to its complex container-style file format standard, there are many kinds of encoding methods for MPEG-4 raw video content and each of them has certain distinction in file structure organization, which will cause the difficulty to locate video content in a uniform way. As a result, we transform the encoded raw content into a standard organization under certain circumstance by the FFMPEG package before loading them into database.

When a time-based MPEG-4 video content request arrived, it will be handled like AVI or MPEG-1 in the flow that described in section 3.1, but they are several differences in step 2 and step 4. For example, in step 2, we will build a 5-Array index

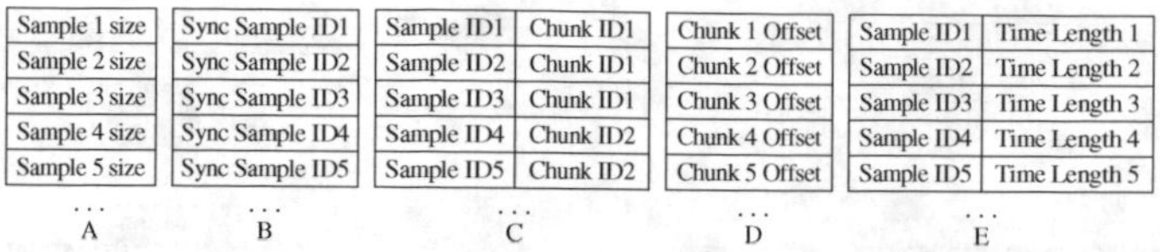

Sample 1 size	Sync Sample ID1	Sample ID1	Chunk ID1	Chunk 1 Offset	Sample ID1	Time Length 1
Sample 2 size	Sync Sample ID2	Sample ID2	Chunk ID1	Chunk 2 Offset	Sample ID2	Time Length 2
Sample 3 size	Sync Sample ID3	Sample ID3	Chunk ID1	Chunk 3 Offset	Sample ID3	Time Length 3
Sample 4 size	Sync Sample ID4	Sample ID4	Chunk ID2	Chunk 4 Offset	Sample ID4	Time Length 4
Sample 5 size	Sync Sample ID5	Sample ID5	Chunk ID2	Chunk 5 Offset	Sample ID5	Time Length 5
...	...	...		...		...
A	B	C		D		E

Fig. 4. Time-based index structure of MPEG-4 file

structure with RLE (Run-Length Encoding) that given in Figure 4; in step 4, the raw content and metadata was reconstructed "chunk by chunk" rather than the way "sample by sample" or "packet by packet'. It should be noticed that chunk is the upper level data structure of sample/packet.

Figure 4 showed that, sample's size, ID and duration time are stored in the A, B and E, respectively. D represented the offset of each chunk, and the array C mapped the relation between each sample and its upper level chunk. The entire illustration of this approach is detailed in our technical report [8].

4 Experimental Results and Analysis

We use the throughput (MB/s) as the primary indicator of performance. In section 4.1, we will list the results of video storage performance and analyze the break-event point of database based video storage. In section 4.2, the performance of our time based video content access approach is presented.

4.1 Break-Event Point of Database Storage

In this section, the detail of insertion and query performance of the 4 storage strategy, as well as the performance impact of database buffer are depicted. In general, the deletion operation in the applications like online video sharing and content based retrieval are performed in the background, it was issued by some staff with a low frequency. Due to its less significance, we only presented it in our technical report [8].

Table 1. Configuration of the test system

CPU	CPU Intel P4 2GHz (Dual-Core)
Main Memory	1GB/4GB
Hard Drive	160GB, 7200 rpm, a volume with 120GB is used for storing data; DBMS A and B are installed on other volume.
OS	Windows 2003 Server Standard Edition with Service Pack1 (NTFS version 3.1)

Our experiments are run at HP PCs, the hardware and software platform are listed it Table 1. The left charts of Figure 5 to Figure 9 are the throughput (*Y-axis*) of A-LOB and A-NTFS, while the right charts are the results of B-LOB and B-NTFS. Their *X-axes* are measured in *storage age*.

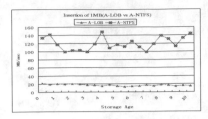 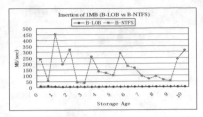

Fig. 5. Insertion performance for 1MB video data

When the video object size is 1MB, the insertion performance is shown in Figure 5, from it we can know the insertion throughput of the two NTFS based storage is superior to the two database based storage, but the performance of B-NTFS is not stable.

Figure 6 showed the query performance of 1MB video objects. In system A, the throughput of NTFS-based storage and LOB-base storage are similar, and A-LOB sometimes is outperforming than the file storage. Figure 6 also showed the different DBMSs BLOB implementation will vary greatly in performance. From the aspects of applications such as content based retrieval, due to most of shots are less than 30 seconds and hence we can say that storing the video objects in system A is valid.

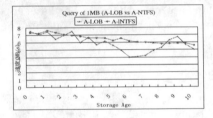 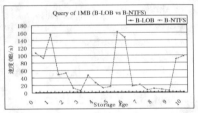

Fig. 6. Query performance for 1MB video data

Figure 7 is the query performance for 512KB video objects. To system A, it is similar to Figure 6, but it is a litter better than Figure 6 in overall where the object size is 1MB. In system B, the NTFS storage is remains outperform than B-LOB overall, but the dominance has decreased greatly.

The superiority of A-LOB where the object size equaled to 256 KB is presented in Figure 8. It showed that A-LOB strategy dominates the performance during all the

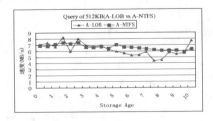 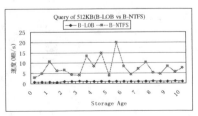

Fig. 7. Query performance for 512KB video data

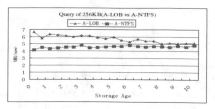
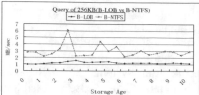

Fig. 8. Query performance for 256KB video data

experiment. As consider to system B, the query performance between B-LOB and B-NTFS also become closed together.

The above results are achieved by using 1GB main memory. We also conduct above evaluation under an HP PC with 4GB main memory, the result (listed in [8]) showed that the main memory didn't have the substantial affection to throughput.

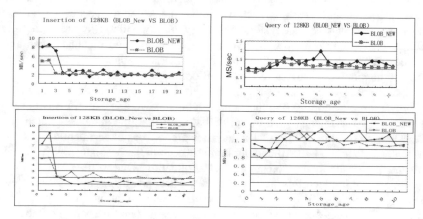

Fig. 9. Buffer configurations have little impact for throughput

According to the experience, we may think that the default settings will impair the performance due to the configuration tuning reasons. Therefore, we conduct experiments to exploit the impacts of some configuration parameters, such as database buffer size and share pool size. Part of results is illustrated in figure 9.

In figure 9, the size of database buffer and share pool in the left top graph and right top graph are 128MB, 0 MB respectively, while the two parameter in the rest two graph is 0 MB and 128MB. From it we can know that, adjusting either database buffer or share pool with abandon the other parameter will take little affection on throughput. However, the role of the combination usage of these two parameters remains under investigation.

The above experimental results of the synthetic workload evaluation indicate that, to the system A, its break-event point is 1MB, where the database based storage works well and have a comparable throughput with NTFS based storage, while the break-event point for system B is roughly around 256KB, it means video objects up to about 256KB are recommended to kept in the database, the larger objects should be placed in the file system for the performance reasons. This phenomenon is conformed to the folklore in principle. It may be interpreted as follows: when the video object is small, compared with query time in database, opening the file for further operation is more time-consuming, which takes a greater proportion in the total time to read and write. In addition, the file system based reads and writes operation may be optimized by the operating system, and the database system will not carry out low-level I/O tuning in default. This is also a possible reason that toward the large video files, the reads and writes performance in file system based storage is better than the performance of database based storage.

4.2 Time-Oriented Video Content Access

We will verify the efficiency of the proposed approach and prove it could be used in practical small video related applications. Our experiments are performed on a PC whose CPU is Pentium IV 2.8G, memory is 4G, and with Windows XP on a disk that is 7200 rpm. We compare the performance/overhead of our database based time-oriented video content access approach with the direct video data access in NTFS file system. To the NTFS file system storage strategy, all raw video files are placed in the same directory within an empty disk volume, their associated metadata such as file path, file name are stored into a relational table in Oracle 10g. To the database based storage, all the video clips are stored in BLOB column in Oracle 10g.

For the simplicity reason, we run the experiments by simulating a client/server style application but running it in the same machine as well as experiments in section 4.1. During the experiments, we use OCI APIs to fulfill all the database-related operations. 5000 high resolution MPEG-1 video files are used as experiment dataset, all of which with a size around 13MB and duration 460s. In the experiments, we take online video sharing website as the potential target application. Due to the duration time of most online shared video clips are shorts from 30 seconds to 2.5 minutes, and few of them longer than 5minutes, we set the performance evaluation strategy simply as follows: Randomly select a video, then issue a request for getting part of content from that video file, the duration of requested content is between 30 seconds and 300 seconds. The average overheads that gained by repeating the above operations is used to compare the performance between the two storage strategies.

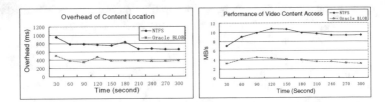

Fig. 10. Overhead of video content access

The overhead of our time-oriented video content approach is divided into two parts: one is the time consumption of calculating and locating the starting offset of segmented video content, the other is cost of fetching the designate content to the client side from the starting offset to ending offset. It should be mentioned that, both strategies cost little at locating the starting offset of designate video clips, and the overhead of our database based approach is far smaller than NTFS based solution. The detail result is shown as the left part of Figure 10. The X-axes and Y-axes in this figure are overhead (millisecond) and the duration of request content (second).

The right part of Figure 10 compared the content access performance for these two solutions. From it we know that the access speed of time-oriented video content access in database is around 3~4.5MB/s, which is slower than file system based strategy, however, it is still much faster than the transmission speeds in personal/home bandwidth network. Considering the applications such as online video sharing or video-on-demand, web transmission speed will become bottleneck of the system under the current network condition. Therefore, it will not affect the video access performance substantially if we stored small video objects into database.

5 Related Work

We have introduced our work on video storage evaluation and database based video content access. In this section, we draw the related literatures from three aspects.

Data Storage: The DB-like system such as BigTable [17] and HBase [4] have been proposed and become popular, both of them are built on top of the distributed file system GFS [2] or HDFS [4] which respond for unstructured raw data storage, e.g. in YouTube system. To the modern DBMSs, all of them are adopt the EXODUS [9] design for efficient insertion or deletion toward a large object by a B-Tree based storage. Besides file system and BLOB based storage, there is another hybrid solution named Data Links [11] which was adopted in DB2. It stores Blobs in the file system, and uses the database to coordinate the BLOB file and its metadata with a transactional semantic consistence.

Storage Performance Evaluation: There are little works concerned with video storage evaluation. SPC-2 [13] is a benchmark paid attentions to read-only on-demand access to video files. The performance study in [12] is measured in the view of long-run and fragmentation analysis. Similarly, the works in [7] studied the large object repository on file system and database under long-run, the fragmentation issues also were considered, further more, it verified the viewpoint indicated by [14] that insertions and deletions within an object can lead to fragmentation.

Video Content Access Approach: At present, most of the video content access approaches are developed in an application case specific way with a file system based data storage. In the modern database system, some commercial DBMS products provided limited functionality to support the video content access, such as Oracle Multimedia [15] and DB2 Video Extender [16]. But both of them only provide the accessibility to some metadata of video file that is still far from enough.

6 Conclusions

Our study on database based video storage revealed the break-event point of large objects storage and verified that the small video clips related applications such as online video sharing are hold a feasibility to adopt the database storage. We also discovered some database buffer configuration which usually play an important role in relational data throughput and found that they have little strength in video data I/O performance. In order to speed up the accessibility of small video objects that suitable stored into database, we devised a time based video content access approach for the video objects in BLOB column of database. It provides a transparent way to wrapper the required content to the end user instead of send the video file from beginning. Our experiments showed that, because of extracted some system-aware metadata and stored them into database in an efficient way, the performance was boosted.

As we known, random and sequential accesses are typical for video files. They naturally require different storage systems: file systems fit sequential accesses and databases for random accesses with indexes. We believe that our initial work roughly indicated a compromise approach for video archive, i.e. store *small* video object into database and the *big* objects are placed in file system. However, there are remain many problems need to be clarified before this hybrid solution become practical.

Acknowledgements

We thank the anonymous reviewers for their valuable comments. This work is supported in part by the HP Lab (China).

References

1. Flynn, R.J., Tetzlatf, W.H.: Multimedia-An introduction. IBM Journal of Research and Development 42(2), 165–176 (1998)
2. Sanjay, G., Howard, G., Shun-Tak, L.: The Google File System. In: Proc. of the 19th ACM Symposium on Operating Systems Principles. Lake George, NY, October 2003, pp. 29–43 (2003)
3. YouTube Architecture,
   http://highscalability.com/youtube-architecture
4. Hadoop Project, http://hadoop.apache.org
5. Now Starring on the Web,
   http://www.wired.com/techbiz/media/news/2006/04/70627
6. Swanson, B.: The Coming Exaflood. Wall Street Journal (January 20, 2007)
7. Sears, R., van Ingen, C.: Fragmentation in Large Object Repositories. In: CIDR 2007, pp. 298–305 (2007)
8. Li, H., Zhou, W., Zhang, X., Wang, S.: Storage strategy of video object and performance evaluation. Technical Report DEKE-TR-2008-1, Key Laboratory of Data Engineering and Knowledge Engineering, Renmin University of China (2008)
9. Carey, M.J., DeWitt, D.J., Richardson, J.E., Shekita, E.J.: Object and File Management in the EXODUS Extensible Database System. In: VLDB (1986)
10. Stonebraker, M., Olson, M.: Large Object Support in POSTGRES. In: ICDE (1993)
11. Bhattacharya, S., Brannon, K.W., Hsiao, H.-I., Mohan, C., Narang, I., Subramanian, M.: Coordinating Backup/Recovery and Data Consistency between Database and File Systems. In: SIGMOD (2002)
12. Seltzer, M., Krinsky, D., Smith, K.A., Zhang, X.: The Case for Application-Specific Benchmarking. In: HotOS (1999)
13. SPC Benchmark-2 (SPC-2) Official Specification, Version 1.2.1

14. Biliris, A.: The Performance of Three Database Storage Structures for Managing Large Objects. In: SIGMOD (1992)
15. Oracle Multimedia:
    `http://www.oracle.com/technology/products/intermedia/`
    `index.html`
16. DB2 Video Extender:
    `http://www-01.ibm.com/software/data/db2/support/`
    `aivextender_z/`
17. Chang, F., Dean, J., Ghemawat, S., Hsieh, W.C., Wallach, D.A., Burrows, M., Chandra, T., Fikes, A., Gruber, R.: Bigtable: A Distributed Storage System for Structured Data. In: OSDI 2006 (2006)

Medical Image Retrieval in Private Network Based on Dynamic Fuzzy Data Model and Wavelet Entropy

Guangming Zhang, Zhiming Cui, and Jian Wu

The Institute of Intelligent Information Processing and Application, Soochow University,
Suzhou 215006, China
gmwell@126.com, szzmcui@suda.edu.cn

Abstract. By analyzing the characters of CT medical image, this paper proposes a new model for CT medical image retrieval, which is using dynamic fuzzy logic and wavelet information entropy. Firstly, the image was decomposed by 2-D discrete wavelet transform to obtain the different level information. Then the entropy from low-frequency subband of CT medical image was calculated, and a membership function based on dynamic fuzzy logic was constructed to adjust the weight for image attribute. At last a model was constructed to obtain the similarity parameter by order for CT image retrieval. The efficiency of our model indicate it is very adaptive to the medical image retrieval in nosocomial private network.

Keywords: image retrieval, wavelet transform, information entropy, dynamic fuzzy logic.

1 Introduction

In recent years, medical image retrieval in nosocomial private network is demanded in many hospitals. Most of the medical images are gray pictures. Due to this character, when we retrieve the medical image we should analysis the similarity in a certain scope. Because a great deal of medical image sequences are taken by the same instrument at different time, so the change trend of similarity could be used in retrieval process. This technique is based on dynamic fuzzy logic. There are a great number of developments and a series of substantial achievements in the domain of fuzzy mathematics' theory research and application, since L. A. Zadeh proposed fuzzy sets in 1965 [1]. However, theses theories can only help to solve those static problems. Dynamic fuzzy logic as an effective theory to solve dynamic fuzzy problems is widely researched. In real world, dynamic fuzzy problems exist universally, especially in the domain of image retrieval. For example, these images become smoother and smoother. The word "become" reflects dynamic character and "smoother" reflects fuzzy character. The whole clause is dynamic fuzzy data. Dynamic fuzzy logic (DFL) based on dynamic fuzzy data is used to solve those problems and has made a series of research achievements. Because of the fuzzy character of the image retrieval, we apply of DFL to analysis the wavelet entropy for medical image retrieval.

Wavelet transform is often used as time-frequency and multiresolution analysis tool, especially in the domain of image processing. In view of the combination wavelet

L. Chen et al. (Eds.): APWeb and WAIM 2009, LNCS 5731, pp. 58–66, 2009.
© Springer-Verlag Berlin Heidelberg 2009

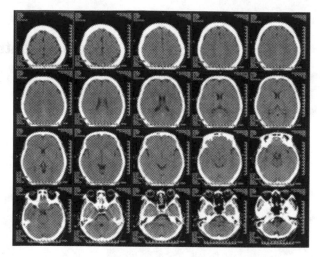

Fig. 1. CT image sequence of human's brain

transform technique and DFL theory, this research is initial. For this reason, an image retrieval based on wavelet transform and DFL is proposed in the paper.

In this paper we use CT image as the object of our research.

The CT image sequence of human's brain is shown in Fig. 1.

We use wavelet transform to get the image entropy in each level, then use DFL to construct a model to analyze the change trend of similarity for medical image retrieval. By experiment, the medical image retrieval using this approach could have the high efficiency for retrieval ratio, both in veracity and speed.

The rest of this paper is organized as follows. The wavelet transform analysis is given in Sections II. Dynamic fuzzy logic theory is given in Sections III. Section IV proposes wavelet information entropy retrieval by DFL. Section V describes our method in the experiment and discusses the results. Conclusions are presented in last section.

2   Wavelet Transform Analysis

Wavelet transform is a multiresolution analysis that represents image variations at different scales [2]. Multi-resolution is an important part of research in several technical domains, including image processing field [3,4]. Consequently, this paper uses the wavelet transform method firstly, which can make a multi-resolution representation where each wavelet coefficient represents the information content of the image at a certain resolution in a certain position.

In real world observed time series are discrete, So discrete wavelet transform (DWT) should be selected for decomposition and reconstruction of time series [5]. There are many DWT algorithms, such as Mallat algorithm [6]. In the paper we use DWT to decompose the CT image.

We define the wavelet series expansion of function $f(x) \in L^2(R)$ relative to wavelet $\psi(x)$ and scaling function $\phi(x)$. The equation can be expressed as follows:

$$f(x) = \sum_k c_{j_0}(k)\phi_{j_0,k}(x) + \sum_{j=j_0}^{\infty}\sum_k d_j(k)\psi_{j,k}(x) \tag{1}$$

where j_0 is an arbitrary starting scale and the $c_{j_0}(k)$'s are normally called the approximation or scaling coefficients, the $d_j(k)$'s are called the detail or wavelet coefficients. The expansion coefficients are calculated as follows:

$$c_{j_0}(k) = \left\langle f(x), \tilde{\phi}_{j_0,k}(x) \right\rangle = \int f(x)\tilde{\phi}_{j_0,k}(x)dx$$
$$d_j(k) = \left\langle f(x), \tilde{\psi}_{j,k}(x) \right\rangle = \int f(x)\tilde{\psi}_{j,k}(x)dx \tag{2}$$

If the function being expanded is a sequence of numbers, like samples of a continuous function $f(x)$. The resulting coefficients are called the discrete wavelet transform of $f(x)$. Then the series expansion defined in following equations becomes the transform pair.

$$W_\phi(j_0,k) = \frac{1}{\sqrt{M}} \sum_{x=0}^{M-1} f(x)\tilde{\phi}_{j_0,k}(x)$$
$$W_\psi(j,k) = \frac{1}{\sqrt{M}} \sum_{x=0}^{M-1} f(x)\tilde{\psi}_{j,k}(x) \tag{3}$$

For $j \geq j_0$, $f(x)$ can be expressed as follows:

$$f(x) = \frac{1}{\sqrt{M}} \sum_k W_\phi(j_0,k)\phi_{j_0,k}(x) + \frac{1}{\sqrt{M}} \sum_{j=j_0}^{\infty}\sum_k W_\psi(j,k)\psi_{j,k}(x) \tag{4}$$

Where $f(x)$, $\phi_{j_0,k}(x)$, and $\psi_{j,k}(x)$ are functions of discrete variable x = 0, 1, 2, ... , M-1.

It is easy to extend the 1-D wavelet transform to the 2-D case [7]. Thus, for an image, the 2-D filter coefficients can be expressed as follows:

$$h_{LL}(m,n) = h(m)h(n), \quad h_{LH}(k,l) = h(k)g(l),$$
$$h_{HL}(m,n) = g(m)h(n), \quad h_{HH}(k,l) = g(k)g(l) \tag{5}$$

The expression where the first and second subscripts denote separately represent the lowpass and highpass filtering along the row and column directions of the image.

For the sake of computation due to the separability of the filters, the wavelet transform can be performed along the rows and columns respectively.

The image could be decomposd as LH which contains horizontal information in high frequency, HL which contains vertical information in high frequency, HH which contains diagonal information in high frequency, and LL which contains the low frequency information as one approximation image. The wavelet transform can decompose the LL band recursively.

The respective coefficients of LH, HL and HH subbands are linearly scaled to the range [0,255] for display. This property is advantageous and is widely used in image analysis.

3   Dynamic Fuzzy Logic (DFL)

In order to apply dynamic fuzzy logic theory in the domain of image retrieval, we introduce the dynamic fuzzy logic concepts, dynamic fuzzy data, then propose estimate of dynamic fuzzy data.

3.1 Dynamic Fuzzy Data

We provided several universal accepted definitions before that the calculation of dynamic fuzzy data is given [8].

Pact 1. The character of data with both dynamic and fuzzy is called dynamic fuzzy character.

Example 1. She becomes more and more beautiful. The word "becomes" reflect the dynamic character and "beautiful" reflect the fuzzy character.

Pact 2. The data with the dynamic fuzzy character is called dynamic fuzzy data.

Example 2. It becomes warmer and warmer. The word "becomes" reflects the dynamic character and "warmer" reflects the fuzzy character. Then we call the whole clause as dynamic fuzzy data.

We based calculation of dynamic fuzzy data on the theory of dynamic fuzzy sets.

Pact 3. We call the collect of dynamic fuzzy data as dynamic fuzzy data sets.

Definition 1. Let a mapping be defined in the domain U.

$$\vec{A}:\vec{a}\to[0,1],\ \vec{a}\to\vec{A}(\vec{a})\ \text{or}\ \overleftarrow{A}:\overleftarrow{a}\to[0,1],\ \overleftarrow{a}\to\overleftarrow{A}(\overleftarrow{a}).$$

We write $(\vec{A},\overleftarrow{A})=\vec{A}$ or $\overleftarrow{A}$, then we named $(\vec{A},\overleftarrow{A})$ the dynamic fuzzy data sets (DFDS) of U ; you can say that is the membership degree of $(\vec{A},\overleftarrow{A})$ to the membership function of $(\overleftarrow{A}(\vec{u})$, $\vec{A}(\vec{u})$). Thus we provided the definition of dynamic fuzzy data sets.

Definition 2. Let(A,A),(B,B)$\in$DF(U), if $\forall(\vec{a},\vec{a})\in$(U), $(\overleftarrow{B},\vec{B})(\vec{a},\vec{a})\leq(\overleftarrow{A},\vec{A})(\vec{a},\vec{a})$ then called $(\vec{B},\vec{B})$ is contained in $(\overleftarrow{A},\vec{A})$ marked as $(\overleftarrow{B},\vec{B})\subseteq(\overleftarrow{A},\vec{A})$ and then $(\overleftarrow{A},\vec{A})=(\overleftarrow{B},\vec{B})$.Obviously, the relation of contain "$\subseteq$" has features as follows:

   (1) Reflexivity: $\forall$(A,A)$\in$DF(U), $(\overleftarrow{A},\vec{A})\subseteq(\overleftarrow{A},\vec{A})$;

   (2) Transitive Characteristic: $(\overleftarrow{A},\vec{A})\subseteq(\overleftarrow{B},\vec{B})$; $(\overleftarrow{B},\vec{B})\subseteq(\overleftarrow{C},\vec{C})\Rightarrow(\overleftarrow{A},\vec{A})\subseteq(\overleftarrow{C},\vec{C})$

   (3) Anti-symmetry: $(\overleftarrow{A},\vec{A})\subseteq(\overleftarrow{B},\vec{B})$; $(\overleftarrow{B},\vec{B})\subseteq(\overleftarrow{A},\vec{A})\Rightarrow(\overleftarrow{A},\vec{A})=(\overleftarrow{B},\vec{B})$

Definition 3. Let (A,A),(B,B)$\in$DF(U), then called $(\overleftarrow{A},\vec{A})\cup(\overleftarrow{B},\vec{B})$ as the union of $(\overleftarrow{A},\vec{A})$ and $(\overleftarrow{B},\vec{B})$, and $(\overleftarrow{A},\vec{A})\cap(\overleftarrow{B},\vec{B})$ as the intersection of $(\overleftarrow{A},\vec{A})$ and $(\overleftarrow{B},\vec{B})$, still we called $(\overleftarrow{A},\vec{A})^c$ as the complementation of $(\overleftarrow{A},\vec{A})$. We provided several following membership functions to the three above calculations:

$$((\overleftarrow{A},\vec{A})\cup(\overleftarrow{B},\vec{B}))(\vec{a},\vec{a})=(\overleftarrow{A},\vec{A})(\vec{a},\vec{a})\vee(\overleftarrow{B},\vec{B})(\vec{a},\vec{a})\overset{\Delta}{=}\max(\overleftarrow{A},\vec{A})(\vec{a},\vec{a}),(\overleftarrow{B},\vec{B})(\vec{a},\vec{a});$$

$$((\bar{A},\vec{A})\cap(\bar{B},\vec{B}))(\bar{a},\vec{a})=(\bar{A},\vec{A})(\bar{a},\vec{a})\wedge(\bar{B},\vec{B})(\bar{a},\vec{a})\triangleq \min(\bar{A},\vec{A})(\bar{a},\vec{a}),(\bar{B},\vec{B})(\bar{a},\vec{a});$$

$$(_{(\bar{A},\vec{A})(a)})^{c}=(1-(\bar{A},\vec{A})(\bar{a},\vec{a}))\triangleq((\bar{1}_-\bar{A}(\bar{a}),\vec{1}_-\vec{A}(\vec{a})).$$

3.2   Estimate of Dynamic Fuzzy Data

Above, we have talked about the calculation of dynamic fuzzy data. In fact, it is another important content to study estimate of dynamic fuzzy data. Here is an example: "A grows faster than B." In this example, we should measure their growth speed. This is the main content in this section. we can defined the estimate of dynamic fuzzy data as:

Definition 4. Mapping: $\mu:\delta\to[\bar{0},\vec{0}]$, $[\bar{1},\vec{1}]$ is called DF estimate, if:

(1) $\mu_{(\bar{\Phi},\vec{\Phi})}=(\bar{0},\vec{0})$, $\mu_{(\bar{X},\vec{X})}=(\bar{1},\vec{1})$

(2) $(\bar{A},\vec{A})\subset(\bar{B},\vec{B})\Rightarrow\mu(\bar{A},\vec{A})\leq\mu(\bar{B},\vec{B})$

(3) $(\bar{A}^n,\vec{A}^n)\uparrow(\downarrow)(\bar{A},\vec{A})\Rightarrow\mu(\bar{A}^n,\vec{A}^n)\uparrow(\downarrow)\mu(\bar{A},\vec{A})$

Then $((\bar{A},\vec{A}),\ \delta,\ \mu)$ is called the space of DF estimate.

Definition 5. If $g_\lambda\to[\bar{0},\vec{0}]$, $[\bar{1},\vec{1}]$ satisfies conditions:

(1) $g_\lambda(\bar{X},\vec{X})=(\bar{1},\vec{1})$;

(2) $g_\lambda((\bar{A},\vec{A})\cup(\bar{B},\vec{B}))=g_\lambda(\bar{A},\vec{A})+g_\lambda(\bar{B},\vec{B})+\lambda g_\lambda(\bar{A},\vec{A})g_\lambda(\bar{B},\vec{B})$, $(\bar{A},\vec{A})\cap(\bar{B},\vec{B})=(\bar{\Phi},\vec{\Phi})$

(3) $(\bar{A}^n,\vec{A}^n)\uparrow(\downarrow)(\bar{A},\vec{A})\Rightarrow\lim\limits_{(\bar{n},\vec{n})+\infty}g_\lambda(\bar{A}^n,\vec{A}^n)=g_\lambda(\bar{A}^n,\vec{A}^n)$

Which is called g_λ estimate.

Theorem 1. While $(\bar{\lambda},\vec{\lambda})>(-\bar{1},-\vec{1})$ then we call g_λ estimate DF estimate.

Theorem 2. $g_\lambda(\bar{\lambda},\vec{\lambda})>(-\bar{1},-\vec{1})$ has features as follows:

(1) $g_\lambda(\bar{\lambda},\vec{\lambda})=\dfrac{(\bar{1},\vec{1})-g_\lambda(\bar{A},\vec{A})}{(\bar{1},\vec{1})+\lambda g_\lambda(\bar{A},\vec{A})}$

(2) $g_\lambda(\bar{A},\vec{A})\cap(\bar{B},\vec{B})=\dfrac{g_\lambda(\bar{A},\vec{A})+g_\lambda(\bar{B},\vec{B})-g_\lambda(\bar{A},\vec{A})(\bar{B},\vec{B})+\lambda g_\lambda(\bar{A},\vec{A})g_\lambda(\bar{B},\vec{B})}{(\bar{1},\vec{1})+\lambda g_\lambda(\bar{A},\vec{A})(\bar{B},\vec{B})}$

4   Wavelet Information Entropy Retrieval by DFL

The goal of image retrieval is to locate the demand image in the certain medical image sequence where store in the database connected by the nosocomial private network. The medical image is contributed to only a few high amplitude coefficients after wavelet transform. Therefore, since the image information is concentrated in a few coefficients, it is possible computer weight of each image to construct retrieval model.

Consequently, a new model which is a quintuple in order to assess the similarity of CT medical image retrieval by DFL is proposed in this paper.

$$EM=(R, N, E, W, S) \tag{6}$$

Where R is a series of CT images that needs to be retrieved, $R=\{R_1,..., R_{n-1},R_n\}$;N is the number of levels of wavelet transform within R; E is the entropy of the certain levels of image after wavelet transformation; W is a set of the weight of each factor of N and E; S is similarity scores set of all the CT images within R, $S=\{S_1,S_2,..., S_n\}$. The final retrieval result is decided by the order of S.

The model has some attributes as follows:

• The number of levels of wavelet transform could be chosen by the precision of retrieval, the large number means high precision but cost more time.
• The entropy of the certain levels of image after wavelet transformation is a objective parameter of this similarity compution.
• We adjust the value of W for the purpose of computer the similarity scores of entropy and levels of wavelet transform levels .Because the value of W is in the interval [0,1], so we could use DFL to construct a membership function to adjust the weight for image retrieval.

We define the domain of discouse $(\overleftarrow{U},\overrightarrow{U})=[\overleftarrow{0},\overrightarrow{0}]$, $[\overleftarrow{1},\overrightarrow{1}]=[0,1]\times[\leftarrow,\rightarrow]$, dynamic fuzzy data sets $(\overleftarrow{A},\overrightarrow{A})$ and $(\overleftarrow{B},\overrightarrow{B})$ indicate "enhancement" and "reduction" respectively. Where "←"denotes increasing or advance direction of dynamic change and "→"indicates reducing or back direction of dynamic change.

For the member of DFD $W \in[0,1]$, then W belongs to DF sets defined as:

$W \underline{DF} (\overleftarrow{w},\overrightarrow{w})=\overleftarrow{w} or \overrightarrow{w}$ and $\max(\overleftarrow{w},\overrightarrow{w})\underline{\underline{\Delta}} \overrightarrow{w}, \min(\overleftarrow{w},\swarrow)\underline{\underline{\Delta}} \overleftarrow{w}$

The calculation between two DFD subsets can be comprehended absolutely as the calculation between its membership function. We construct the membership function as follows:

$$\overleftarrow{A}(\overleftarrow{u}) = \begin{cases} 0 & \text{if } 0 \leq \overleftarrow{u} \leq \overleftarrow{0.5} \\ (1+(\dfrac{\overleftarrow{u}-0.5}{0.05})^{-2})^{-1} & \text{if } \overleftarrow{0.5} \leq \overleftarrow{u} \leq \overleftarrow{1} \end{cases}$$

$$\overrightarrow{A}(\overrightarrow{u}) = \begin{cases} 0 & \text{if } 0 \leq \overrightarrow{u} \leq \overrightarrow{0.5} \\ (1+(\dfrac{\overrightarrow{u}-0.5}{0.05})^{-2})^{-1} & \text{if } \overrightarrow{0.5} \leq \overrightarrow{u} \leq \overrightarrow{1} \end{cases}$$

$$\overleftarrow{B}(\overleftarrow{u}) = \begin{cases} 0 & \text{if } 0 \leq \overleftarrow{u} \leq \overleftarrow{0.5} \\ (1+(\dfrac{\overleftarrow{u}-0.5}{0.05})^{-2})^{-1} & \text{if } \overleftarrow{0.5} \leq \overleftarrow{u} \leq \overleftarrow{1} \end{cases}$$

$$\vec{B}(\vec{u}) = \begin{cases} 0 & \text{if } 0 \leq \vec{u} \leq \overrightarrow{0.5} \\ (1+(\dfrac{\vec{u}-0.5}{0.05})^{-2})^{-1} & \text{if } \overrightarrow{0.5} \leq \vec{u} \leq \vec{1} \end{cases}$$

(7)

By using the constructed membership function of DFL, we could find the trend of similarity, so the performance of retrieval based on DFL is prominent.

5   Experimental Results and Discussion

Before the experiment, we get a lot of CT images from the First Affiliated Hospital of Soochow University, including the brain's images and DSA images.

5.1   The Experimental Using DFL

In order to retrieve the CT medical image, we proposed a new model combine 2-D discrete wavelet transform and DFL. The steps of this processing method as follows:

• All CT images are decomposed into low-frequency subband and high-frequency subband based 2-D discrete wavelet transform, the wavelet filter used was Haar and decomposition level was restricted to three, after initial experimentation with 1 to 3 decomposition levels.
 • Calculate the entropy from low-frequency subband of CT medical image.
 • Construct a model based on DFL to obtain the similarity degree in order.

The steps we presented are illustrated as follows:

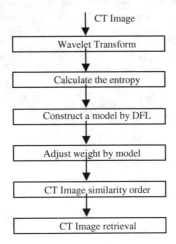

Fig. 2. Image retrieval using DFL and wavelet entropy

5.2   The Effect Analysis

Due to the disordered distributing of the CT medical image library with a mass of information, most medical information could not be utilized and accessed. In order to

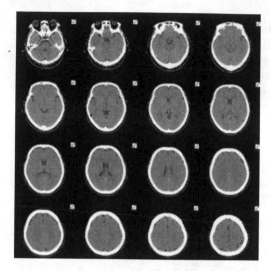

Fig. 3. CT image retrieval result

Table 1. Data analysis

R	Number of levels	Entropy	weight	similarity
1	3	3.026	0.75	100.00%
2	2	3.117	0.85	82.58%
3	3	3.263	0.70	81.23%
4	3	3.285	0.68	78.41%
5	2	3.312	0.82	75.45%

lookup certain image quickly and accurately and make best use of these data, the CT medical images retrieval technology are applied to adjuvant therapy and surgical planning purposes.

In the experiment, the CT image retrieval result is shown in Fig. 3. Use this paper's arithmetic, the result similarity with the descending order is shows in the Table 1.

6 Conclusions

In order to retrieve CT medical image, a new model which is using dynamic fuzzy logic and wavelet information Entropy was proposed. Firstly, the image was decomposed by discrete wavelet transform to obtain the different subband information. Then we calculate the entropy from low-frequency subband of CT medical image, and constructed a membership function based on dynamic fuzzy logic to adjust the weight for image attribute. At last we construct a model to obtain the similarity parameter by order in CT image retrieval. The efficiency of our model indicate it is very fit for the medical image retrieval in hospital private network. Further investigations on the retrieval different formats of medical image are left for future work.

Acknowledgement

This research was partially supported by the Natural Science Foundation of China (No.60673092), the Higher Technological Support and Innovation Program of Jiangsu Province in 2008 (BE2008044) and the Higher Education Graduate Research Innovation Program of Jiangsu Province in 2008 (CX08B_099z).

References

1. Zadeh, L.A.: Fuzzy sets. J. Information and Control 8, 338–353 (1965)
2. Mallat, S.: A Wavelet Tour of Signal Processing. Academic Press, Boston Mass (1998)
3. Usevitch, B.E.: A Tutorial on Modern Lossy Wavelet Image Compression: Foundations of JPEG 2000. IEEE Signal Processing Magazine 18, 22–23 (2001)
4. Rajpoot, K., Rajpoot, N., Noble, J.A.: Discrete wavelet diffusion for image denoising. In: Elmoataz, A., Lezoray, O., Nouboud, F., Mammass, D. (eds.) ICISP 2008. LNCS, vol. 5099, pp. 20–28. Springer, Heidelberg (2008)
5. Gonzolez, R.C., Woods, R.E.: Digital Image Processing, 2nd edn. Prentice Hall, Englewood Cliffs (2002)
6. Yang, X., Yang, W., Pei, J.: Different focus points images fusion based on wavelet decomposition. In: Proceedings of Third International Conference on Information Fusion, vol. 1, pp. 3–8 (2000)
7. Alexandera, M.E., Baumgartnera, R., Windischbergerb, C., Moserb, E., Somorjaia, R.L.: Wavelet domain de-noising of time-courses in MR image sequences. J. Magnetic Resonance Imaging 18, 1129–1134 (2000)
8. Fanzhang, L.: Dynamic Fuzzy Logic and Its Applications. Nova Science Pub Inc., New York (2008)

Extension of OWL with Dynamic Fuzzy Logic

Zhiming Cui, Wei Fang, Xuefeng Xian, Shukui Zhang, and Pengpeng Zhao

Jiangsu Provincial Key Laboratory of Computer Information Processing Technology,
Soochow University, Suzhou
The Institute of Intelligent Information Processing and Application,
Soochow University, Suzhou, 215006, P.R. China
064027065001@suda.edu.cn

Abstract. In recent years, ontology has played a major role in knowledge representation. Ontology languages are based on description logics. Though they are expressive enough, they cannot express and reason with fuzzy and dynamic knowledge on the Semantic Web. To deal with uncertain and dynamic knowledge on the Semantic Web and its applications, a new fuzzy extension of description logics,OWL and Ontology based on Dynamic fuzzy logic called the dynamic Description logics(DFDL), dynamic fuzzy Ontology(DFO) and dynamic fuzzy OWL (DFOWL) are presented. The syntax and semantics of DFDL, DFO and DFOWL are formally defined, and the forms of axioms and assertions are specified. The research indicates the DFOWL provides more expressive power for the Semantic Web, and overcomes the insufficiency of OWL as the ontology language for the Semantic Web.

Keywords: dynamic fuzzy logic; semantic web; ontology; OWL.

1 Introduction

In recent years, Ontology has played a major role in knowledge representation for the Semantic Web. Ontology is a conceptualization of a domain into a human understandable, and machine-readable or machine-processable format consisting of entities, attributes, relationships, and axioms[1].The OWL(Web Ontology Language) is designed for use by applications that need to process the content of information instead of just presenting information to humans. And the OWL is intended to provide a language that can be used to describe the classes and relations between them that are inherent in Web documents and applications. OWL facilitates greater machine interpretability of Web content than that supported by XML, RDF, and RDFS by providing additional vocabulary along with a formal semantics[2].

The Semantic Web[3] is a vision for the future of the Web in which information is given explicit meaning, making it easier for machines to automatically process and integrate information available on the Web. Description logics (DLs)[4] are widely used on the semantic web. Fuzzy extensions of description logics import the fuzzy theory to enable the capability of dealing with fuzzy knowledge[3]. The fuzzy knowledge plays an important role in many domains that face a huge

L. Chen et al. (Eds.): APWeb and WAIM 2009, LNCS 5731, pp. 67–76, 2009.

amount of imprecise and vague knowledge and information, such as text mining, machine learning, information integration and natural language processing[5]. On the Semantic Web, the knowledge expression is a very key problem, but a lot of knowledge has a dynamic and fuzzy characters, traditional logical method or fuzzy logic or some other knowledge expression models are very difficult to express them accurately and effectively, such as, She is a girl who becomes more and more beautiful. Here " become", "beautiful" have embodied " dynamic character " and "fuzzy character" sufficiently. If we use the existing approaches to resolve these problems having dynamic and fuzzy characters will be very difficult, they can only represent static knowledge.

In addition, SHOIN(D) is the theoretical counterpart of the OWL Description Logic[6]. Thus, in the paper, we define a dynamic fuzzy extension of the OWL language considering fuzzy SHOIN(D). We have extended the syntax and semantic of fuzzy SHOIN(D) with the possibility and dynamic to add a concept modifier to a relationship and introducing a novel constructor which enables us to define a subset of concepts with a membership value greater or lower that a fixed value. The main contribution of this paper is the description of how we transfer a classical DL , ontology and OWL to dynamic fuzzy description logic (DFDL), dynamic fuzzy ontology (DFO) and dynamic fuzzy OWL (DFOWL) which have better representation and inference ability for fuzzy and dynamic knowledge on the Semantic Web.

2   Dynamic Fuzzy Description Logics

A lot of information on the Semantic Web[7][3] are uncertain, imprecise and dynamic. And, the traditional Description Logics cannot express and inference these dynamic knowledge efficiently. Then, to deal with these uncertain and dynamic knowledge on the Semantic Web, we have extended the Fuzzy Set and Description Logic based on Dynamic Fuzzy Logic(DFL)[8][9].

2.1   Dynamic Fuzzy Logic

Definition 1. A statement having character of dynamic fuzzy is called dynamic fuzzy proposition that is usually symbolized by capital letters A, B, C... . E.g.1 Here is a DF proposition: It will be getting hot soon.

Definition 2. A dynamic fuzzy number $(\overleftarrow{a}, \overrightarrow{a}) \in [0,1]$,which is used to measure a dynamic fuzzy proposition's true or false degree, is called dynamic fuzzy proposition's true or false. It is usually symbolized by $(\overleftarrow{a}, \overrightarrow{a})$, $(\overleftarrow{b}, \overrightarrow{b})$, $(\overleftarrow{c}, \overrightarrow{c})$...., where $(\overleftarrow{a}, \overrightarrow{a}) = \overleftarrow{a}$ or $\overrightarrow{a}$, $\min(\overleftarrow{a}, \overrightarrow{a}) \overset{\Delta}{=} \overleftarrow{a}$, $\max(\overleftarrow{a}, \overrightarrow{a}) \overset{\Delta}{=} \overrightarrow{a}$, the same are as follows.

Definition 3. A dynamic fuzzy proposition can be regarded as a variable whose value is in the interval $[0,1] \times [\leftarrow, \rightarrow]$. The variable is called dynamic fuzzy proposition variable that is usually symbolized by small letter.

Operation rules of any dynamic fuzzy variable $(\overleftarrow{x},\overrightarrow{x}),(\overleftarrow{y},\overrightarrow{y}) \in[0,1]$ are prescribed as follows:

①Negation "–"

The negation of variable $(\overleftarrow{x},\overrightarrow{x})$ is presented by $\overline{(\overleftarrow{x},\overrightarrow{x})}$, and $\overline{(\overleftarrow{x},\overrightarrow{x})} \overset{\Delta}{=}((1-\overleftarrow{x},1-\overrightarrow{x}))$

②Disjunction "∨"

$(\overleftarrow{x},\overrightarrow{x}) \vee (\overleftarrow{y},\overrightarrow{y}) \overset{\Delta}{=}\max((\overleftarrow{x},\overrightarrow{x}),(\overleftarrow{y},\overrightarrow{y}))$

③Conjunction "∧"

$(\overleftarrow{x},\overrightarrow{x}) \wedge (\overleftarrow{y},\overrightarrow{y}) \overset{\Delta}{=}\min((\overleftarrow{x},\overrightarrow{x}),(\overleftarrow{y},\overrightarrow{y}))$

④Condition "→"

$(\overleftarrow{x},\overrightarrow{x}) \to (\overleftarrow{y},\overrightarrow{y}) \Leftrightarrow \overline{(\overleftarrow{x},\overrightarrow{x})} \vee (\overleftarrow{y},\overrightarrow{y}) \overset{\Delta}{=}\max(\overline{(\overleftarrow{x},\overrightarrow{x})},(\overleftarrow{y},\overrightarrow{y}))$

⑤bi-direction "↔"

$(\overleftarrow{x},\overrightarrow{x}) \leftrightarrow (\overleftarrow{y},\overrightarrow{y}) \overset{\Delta}{=}\min(\max(\overline{(\overleftarrow{x},\overrightarrow{x})},(\overleftarrow{y},\overrightarrow{y})),\max(\overline{(\overleftarrow{x},\overrightarrow{x})},(\overleftarrow{y},\overrightarrow{y})))$

Definition 4. Dynamic fuzzy calculus formations can be defined as follows:

(1) A simple dynamic fuzzy variable itself is a well-formed formula.

(2) If $(\overleftarrow{x},\overrightarrow{x})P$ is a well-formed formula, $\overline{(\overleftarrow{x},\overrightarrow{x})P}$ is a well-formed formula, too.

(3) If $(\overleftarrow{x},\overrightarrow{x})P$ and $(\overleftarrow{y},\overrightarrow{y})Q$, are well-formed formulas, $(\overleftarrow{x},\overrightarrow{x})P\vee(\overleftarrow{y},\overrightarrow{y})Q$, $(\overleftarrow{x},\overrightarrow{x})P\wedge(\overleftarrow{y},\overrightarrow{y})Q$, $(\overleftarrow{x},\overrightarrow{x})P\to(\overleftarrow{y},\overrightarrow{y})Q$, $(\overleftarrow{x},\overrightarrow{x})P\leftrightarrow(\overleftarrow{y},\overrightarrow{y})Q$ are also well-formed formulas.

(4) A string of symbols including proposition variable connective and brackets is well-formed formula if and only if the strings can be obtained in a finite of steps, each of which only applies the earlier rules (1),(2) and (3).
The main formulas can be found in reference[8][9].

2.2 Dynamic Fuzzy Description Logics

The fuzzy description logic(FDL or FALC)[13][14][15] interpret concepts or roles as fuzzy sets of individuals or individual pairs. Such concepts and roles are called fuzzy concepts and fuzzy roles. But the Fuzzy Description Logics (FDL) cannot express the dynamic knowledge. Then FDL is extended with dynamic fuzzy capabilities to yield DFDL (Dynamic Fuzzy Description Logics) in which concepts are interpreted as dynamic fuzzy sets. For example , in DFDL, a concept C is interpreted as a dynamic fuzzy set and a statement like " a is C " has a truth-value in $[0,1]\times[\leftarrow,\to]$. In this case, $\cdot^{I(s)}$ is an interpretation function mapping C into a membership function $C^{I(s)}, C^{I(s)}:\Delta^{I(s)} \to [0,1]\times[\leftarrow,\to]$. S is state of DFDL. Acting on concepts, the crisp operations of conjunction, disjunction, negation and quantification are normally extended to their dynamic fuzzy counterparts. By using the dynamic fuzzy logic, we can say it is becoming hotter and hotter to a degree of $\overrightarrow{0.17}$, another day is hot to a degree of $\overleftarrow{0.19}$.

DFDL are a set N_c of concept names and a set N_R of role names. The syntax of DFDL is the extending syntax combined FDL with DFL.

Concepts and relationship C,D of DFDL are defined with the following syntax rules:

(1) $C, D \rightarrow \neg C | C \cap D | C \cup D | \exists R.C | \forall R.C | < \alpha > C | < \alpha > D$

(2) $C, C^{I(s)} \rightarrow < \alpha > C | < \alpha > C^{I(s)}$

(3) $R \rightarrow r | < \alpha > R$

Where $C_i \in N_c$, $R \in N_R$, α is an action $[0,1] \times [\leftarrow, \rightarrow]$, r is an atom relationship.

Definition 5. Recursive formula definition of DFDL predicate formation is as follows:

(1) An atom (first order logical symbols) is a formula.

(2) Assertion formula and general formula are both formula.

(3) If G and H are formulas, T is a dynamic fuzzy truth value of assignment, $(\overleftarrow{x}, \overrightarrow{x})$ is a free variable in DFDL, $\overline{G}, G \vee H, G \wedge H, G \rightarrow H$, $G \leftrightarrow H$, $(\overleftarrow{x}, \overrightarrow{x})G$, $\forall(\overleftarrow{x}, \overrightarrow{x})G$, $\exists(\overleftarrow{x}, \overrightarrow{x})G$ are all formulas.

(4) Any string of symbol is a formula of DFDL, if and only if the string can be obtained in a finite of steps, each of which only applies the earlier rules (1), (2) and (3).

The assertion formula of DFDL is as follows:

Dynamic Fuzzy TBox DFTB. *A* dynamic fuzzy TBox DFTB consists of a finite set of dynamic fuzzy concept inclusion axioms of the form $< c \geq n >, < c \leq n >, < c > n >, \alpha \subseteq \beta$, and $\alpha \equiv \beta$, where c is a concept including axiom, α and β is action.

Dynamic Fuzzy ABox DFAB. *A* dynamic fuzzy ABox DFAB consist of a finite set of dynamic fuzzy concept and fuzzy role assertion axioms of the form $< a \geq n >, < a \leq n >$, $R(a,b), < \alpha > C(a)$, and $< \alpha > C(a,b)$, Where a, b is a concept or role assertion.

Dynamic Fuzzy Knowledge Base DFKB. *A* dynamic fuzzy knowledge base DFKB= <DFTB,DFAB> consists of a dynamic TBox DFTB, and a dynamic fuzzy ABox DFAB.

Definition 6. If the knowledge base DFKB= <DFTB,DFAB> of DFDL C is satisfied with the equation: W |=DFTB, and W |=DFAB, then the DFKB is satisfiable, which denoted by W |=DFKB.

2.3   Semantic Interpretation of DFDL

In DFDL, s is a state of DFDL. The fuzzy interpretation of s is defined as mapping $I = < \Delta^{I(s)}, \cdot^{I(s)} >$, where $\Delta^{I(s)}$ is a nonempty set as the domain. Mapping function $\cdot^{I(s)}$ makes a concept map to a dynamic fuzzy subset of $\Delta^{I(s)}$, and let relationship map a subset of $\Delta^{I(s)} \times \Delta^{I(s)}$. The semantics of DFDL is extended. The main idea is that concepts and roles are interpreted as fuzzy subsets of an interpretation domain. Therefore, axioms, rather being satisfied (true) or unsatisfied (false) in an interpretation, become a degree of truth in $[0,1] \times [\leftarrow, \rightarrow]$.

In general, according to construction operators and description logic construct the complex concept and simple concepts relations. Then DFDL at least contain the following operators: Conjunction $\cap$, Disjunction $\cup$, Not $\neg$, Existential quantification $\exists$, and Value restriction $\forall$. DFDL combined with the basis of the time constraints operator α $[\leftarrow,\rightarrow]$ and FDL will extend to DFDL. So, the syntax and semantic of DFDL are shown in table 1 and table 2.

Table 1. Interpretation of concepts in DFDL

DFDL	Syntax	Semantic
Concept	A	$A^{I(s)} \subseteq \Delta^{I(s)}$
Role name	R	$R^{I(s)} \subseteq \Delta^{I(s)} \times \Delta^{I(s)}$
Conjunction	$C \cap D$	$C^{I(s)} \cap D^{I(s)}$
Disjunction	$C \cup D$	$C^{I(s)} \cup D^{I(s)}$
Value restriction	$\forall R.C$	$\{x \in \Delta^{I(s)} | \forall y \cdot (x,y) \in R^{I(s)} \Rightarrow y \in C^{I(s)}\}$
Existential quantification	$\exists R.C$	$\{x \in \Delta^{I(s)} \exists y \cdot (x,y) \in R^{I(s)}\}$
Top	$\top$	$\Delta^{I(s)}$
Bottom	$\perp$	Φ
Negation	$\neg C$	$\Delta^{I(s)} - C^{I(s)}$

Table 2. Interpretation of axioms in DFDL

Abstract Syntax	Syntax	Semantic
Class(A partial $C_1,\ldots,C_n$)	$A \subseteq C_1 \cap \ldots \cap C_n$	$A^{I(s)} \subseteq C_1^{I(s)} \cap \ldots \cap C_n^{I(s)}$
Class(A complete $C_1,\ldots,C_n$)	$A = C_1 \cap \ldots \cap C_n$	$A^{I(s)} = C_1^{I(s)} \cap \ldots \cap C_n^{I(s}$
SubClassOf(C_1 ,C_2)	$C_1 \subseteq C_2$	$C_1^{I(s)} \subseteq C_2^{I(s)}$
Transitive(R)	$Tr(R)$	$R^{I(s)} = (R^{I(s)})^+$
SubPropertyOf(P_1 ,P_2)	$P_1 \subseteq P_2$	$P_1^{I(s)} \subseteq P_2^{I(s)}$
Valued(R_1,o_1)$\ldots$ value(R_n,o_n)	$(o,o_i):R_i$	$(o^{I(s)},o_i^{I(s)}) \in R_i^{I(s)}$
SameIndividual($o_1,\ldots, o_n$)	$o_1 = o_2 = \ldots = o_n$	$o_1^{I(s)} = o_2^{I(s)} = \ldots = o_n^{I(s)}$
DifferntIndividuals($o_1,\ldots, o_n$)	$o_i \neq o_i, i \neq j$	$o_i^{I(s)} \neq o_{i,}^{I(s)}, i \neq j$

The assertions of DFDL is: $<a:Con>$, $I|=< a : C \geq n >$ iff $C^{I(s)} (a^{I(s)}) \geq n$, and the terminological axioms is A=C or $A \subseteq CI|=A \subseteq C$ iff $\forall x \in^{I(s)} . A^{I(s)} (x) \leq C^{I(s)}$.

Definition 7. A rule of Dynamic Fuzzy Production can be defined as follows:
$P \leftarrow Q$, CDF, I. Its right hand side I is a group of conditions, left hand side is some actions. Premise Q and conclusion P both may be DF. CDF is $(\overleftarrow{0},\overrightarrow{0}) \leq$CDF$\leq (\overleftarrow{1},\overrightarrow{1})$, be called Degree of Confidence.

When $Q = a_1(u_1)a_2(u_2)\ldots a_n u_n$,if $Q^1 = a_1^1(v_1) a_2^1(v_2)\ldots a_n^1(v_n)$. t^1 (herein t^1:$0< t^1 \leq 1$),and $a^1 = a$or *,i=1,2, $\ldots$,n, $0 \leq u_i \leq v_i \leq 1$ $(0 \leq v_i \leq u_i \leq 1)$,i= 1,2,$\ldots$,n then Q is Matching. Here the distance of Q and Q^1 can be defined as follows:

$d=\max\{|v_1 - u_1|, |v_2 - u_2 \quad |, \ldots, |v_n - u_n|\}$
thus the definition of matching degree of premise Q is:
$m=\min\{1\text{-}d, t^1\}$ (or $m = t^1 *(1\text{-}d)$,etc)

3   Dynamic Fuzzy Ontology

In this section, we formally describe the notion of dynamic Fuzzy Ontology. Ontology provide a criterion way to represent and share knowledge using a common vocabulary between human and computer. Ontology is an effective conceptualism commonly used for the Semantic Web. Ontology languages are based on Description Logics (DLs). Informally, an ontology consists of a hierarchical description of important concepts in a particular domain, along with the description of the properties (or the slots and the instances) of each concept. DLs play a key role in this context as they are essentially the theoretical counterpart of OWL DL.

Fuzzy ontology is capable of modeling fuzzy set through representing the fuzzy term and membership functions with rules[6]. So in theory it is possible to apply such dynamic fuzzy ontology into Semantic Web [7][12]. To handle that, dynamic fuzzy ontology are defined to reflect the "human-like" vague and dynamic knowledge. Dynamic fuzzy ontology (DFO) is ontology that evolve in time to adapt to the environment what they need. Generally, a dynamic fuzzy ontology structure can be defined as consisting of concepts, of dynamic fuzzy relations among concepts associative relationships and of a set of ontology axioms, expressed in an appropriate logical language. So, a dynamic Fuzzy Ontology structure according to the reference[11] can be defined as follows.

Definition 8. A dynamic Fuzzy Ontology structure is a sextuple O := (C, R, T, NT, A, DF), where C is a set of (dynamic fuzzy) concepts (or classes , in OWL — of individuals, or categories), i.e., wine, book, animal, plane , etc. R is set of (dynamic fuzzy) relations(or roles, or slots) in C×C, i.e., the concept hand has a part-of relationship with the concept person. T is a relation in C×C, called Taxonomy, i.e., apple is_a fruit. NT is a set of non-taxonomic(fuzzy)associative relationships that relate concepts across tree structures, i.e., Naming relationships, Locating relationships, Functional relationships, etc. A is a set of ontology axioms(or rules), expressed in an appropriate logical language, i.e., asserting class subsumption, equivalence, individuals, relationships and functions, etc. DF is a constrained condition denoted by [0,1] ×[←,→]. That is to say, a dynamic Fuzzy Ontology is an ontology extended with dynamic fuzzy values assigned through the functions:

   i: Instance ↦[0,1] ×[←,→]
   v: Property_values ↦[0,1] ×[←,→]

The relationship between concepts in a dynamic fuzzy ontology is provided with a dynamic fuzzy value: rel: Instances×Instances↦[0,1] ×[←,→].

The semantics have been extended. The main idea is that concepts and roles are interpreted as fuzzy subsets of an interpretation's domain[19]. Therefore, axioms, rather being satisfied (true) or unsatisfied (false) in an interpretation, become a degree of truth in [0,1] ×[←,→].

Dynamic Fuzzy Operator(DF) is denoted by$[\leftarrow,\rightarrow]$. It means that the current state in certain time has two change directions: "$\rightarrow$"means the good or advance direction, but "$\leftarrow$"means the bad or back direction.

Dynamic Fuzzy Semantics(DFSS) describes the meanings of sentence structures with the characters of dynamic and fuzzy[9]. DFSS is composed of a group of dynamic fuzzy rules (or axioms) whose basic elements are configurations denoted by $<s,\delta, [\leftarrow,\rightarrow]>$, where $[\leftarrow,\rightarrow]$ is symbolized by dynamic fuzzy operator. $<s, \delta, [\leftarrow,\rightarrow]>$ acts on the current state δ and the action waiting for executing is s. In our case s is denoted by state, $\leftarrow$is symbolized by state set, $s((\overleftarrow{x},\overrightarrow{x}))$ is indicated by the value or content of the variable$(\overleftarrow{x},\overrightarrow{x})$ in the state s,$< (\overleftarrow{x},\overrightarrow{x})$,s$>$ stands for the variable$(\overleftarrow{x},\overrightarrow{x})$ waiting for evaluating in the state s, $\delta[(\overleftarrow{n},\overrightarrow{n})(\overleftarrow{d},\overrightarrow{d})/(\overleftarrow{x},\overrightarrow{x})]$ means the state where $(\overleftarrow{n},\overrightarrow{n})(\overleftarrow{d},\overrightarrow{d})$ takes the place of $(\overleftarrow{x},\overrightarrow{x})$ in the state s.

For example, we use A describe the fact "It will rain". Then we can denote as $(\overleftarrow{A},\overleftarrow{0.2})$, where $\overleftarrow{0.2}$ indicates the light rain getting stop and 0.2 which denotes light rain here shows the degree of rain. Of course, since fuzzy sets are a sound extension of classical boolean sets, it is always possible to define crisp (i.e, non-fuzzy) concepts (resp., relations) by using only values in the set $[0,1] \times[\leftarrow,\rightarrow]$.

4 Dynamic Fuzzy OWL

Once we have defined a dynamic fuzzy ontology (DFO) and after have shown how to extend OWL(such as SHOIN(D)), the next step is to define the new dynamic fuzzy language suitable to implement the dynamic fuzzy ontology. OWL[12][11]is an ontology language that has recently been a W3C recommendation. OWL mainly describes the objective world from two aspects, concept and attribute. The corresponding measure of description is object-oriented domain and oriented data type domain. And the concept of OWL expresses with classes, with a name (such as URI) or expression, moreover, providing a great of to build expression. The power expression ability is exactly decided by its concept , character constructor, and all kinds of axioms. KAnon's ontology language is based on RDFS[16] with proprietary extensions for algebraic property characterizations, cardinality, modularization, meta-modeling and explicit representation of lexical information[6]. However, OWL DL does not deal with the knowledge represented with a vague and dynamic definition. A new extension of OWL DL, named DFOWL, by adding a dynamic fuzzy value to the entities and relationships of the ontology following the dynamic fuzzy logic syntax of Section 2 and the dynamic fuzzy ontology definition given in Section 3.

Formally, a dynamic fuzzy set A with respect to a universe X is characterized by a membership function $\mu A:X\rightarrow[0,1]\times[\leftarrow,\rightarrow]$, assigning an A-membership degree, $\mu A(x)$, to each element x in X. $\mu A(x)$gives us an estimation of the belonging x to A. Typically, if $\mu A(x)=1$ then x definitely belongs to A, while $\mu A(x)= \overrightarrow{0.8}$ means that x is "likely" develop to be a term of A.

For example, we can represent stock market with the following RDF/XML syntax in dynamic fuzzy OWL's DL language.

```
<owl:Class rdf:ID="Stock">
<owl:unionOf rdf:parseType="Collection">
<owl:Class rdf:about="#000001"
Type="→" fuzzy:degree="0.8"/>
<owl:Class rdf:about="#000002"
Type="←" fuzzy:degree="0.75"/>
< /owl:unionof>
< /owl:Class>
```

In general, according to construction operators the description logic construct the complex concept and relation on simple concepts and relations. Description logic is usually at least contain the following operators: Conjunction $\sqcap$, Disjunction $\sqcup$, Not $\neg$, Existential quantification $\exists$, and Value restriction $\forall$. The basic OWL is added on the basis of the time constraints operator($\overleftarrow{x}, \overrightarrow{x}$), will form DFOWL.

A dynamic fuzzy interpretation I with respect to a concrete domain D is a pair $I=(\Delta^{I(s)}, \cdot^{I(s)})$ consisting of a non empty set $\Delta^{I(s)}$ (called the domain at the state s),disjoint from Δ_D ,where $\Delta^{I(s)}$ is the domain of interpretation, as in the classical case, and $\Delta \cdot^{I(s)}$ is an interpretation function which maps concepts (roles) to a membership function $\Delta^{I(s)} \rightarrow [0,1] \times [\leftarrow, \rightarrow] (\Delta^{I(s)} \times \Delta^{I(s)} \rightarrow [0,1] \times [\leftarrow, \rightarrow])$, which defines the dynamic fuzzy subset $C^{I(s)}$ $(R^{I(s)})$. To each $c \in I_c$ an element in $_D$, to each $T \in R_c$ a subset of and to each n-array concrete predicate d the interpretation $d^D \subseteq \Delta_D^n$.

A dynamic fuzzy concept C is satisfiable iff there exists some dynamic fuzzy interpretation I for which there is some $a \in \Delta^{I(s)}$ such that $C^{I(s)}$ $(a) = n$, and $n \in [0,1] \times [\leftarrow, \rightarrow]$, A dynamic fuzzy interpretation I satisfies a TBOX Tiff $\forall a \in \Delta^{I(s)} R^{I(s)} \leq D^{I(s)}$, for each $R \subseteq C$, and $\forall a \in \Delta^{I(s)} R^{I(s)} = D^{I(s)}$, for each $R \equiv C$.

5   Related Work

Nowadays, on the Semantic Web quite a lot of ontology languages exist,like the OWL and DAML+OIL. Both these languages, use Description logic(Dls) as their underlying formal for representation of knowledge as well as for performing tasks. Fuzzy sets theory, introduced by L. A. Zadeh [10], allows to deal with imprecise and vague data, so that a possible solution is to incorporate fuzzy logic into ontologies. In " On the Expressiveness of the Languages for the Semantic Web - Making a Case for 'A Little More'", Ch. Thomas and A. Sheth introduce the need for fuzzy probabilistic formalisms on the Semantic Web, in particular within OWL. In "Fuzzy ontologies for information retrieval on the WWW", D. Parry uses fuzzy ontologies, and presents a broad survey of relevant techniques, leading up to the notions of fuzzy search and fuzzy ontologies. Handling faceted or vague information is an open issue in many research areas, for example as in object-oriented databases systems[15]. Also the conceptual formalism supported by a typical ontology[9] may not be sufficient to represent uncertain information that is commonly found in many application domains. In addition, the way in which concepts and relations are usually expressed can be inadequate to handle the nuances of natural languages used by human to describe and to understand the context in which they live. The fuzzy set theory [4], originally introduced by L. A. Zadeh [10], allows one to denote non-crisp concepts. Fuzzy sets and ontologies have been jointly used to resolve uncertain information problems in various areas, for example, in text retrieval [17][1]or to generate a scholarly ontology from a database in the ESKIMO [2] and FOGA [6] frameworks. However, there is not a complete fusion of fuzzy set theory with ontologies in any of these examples. Kang et al.[15] presented a description logics for fuzzy ontologies on semantic web.

6   Conclusion

"The success of the deployment of the Semantic Web will largely depend on whether useful ontologies will emerge, allowing shared agreements about vocabularies for knowledge representation." [14] In this paper we have extended the DL language, ontology and OWL with dynamic fuzzy logic theory. We introduce the DFL theory like semantic interpretation and reasoning rules The combination of transitive and inverse roles allow us encode and reason with fuzzy and dynamic knowledge on the Semantic Web. Furthermore ,the incorporation of DFL allows us to encode and reason with imprecise and dynamic knowledge. More representation features and efficient reasoning and analysis algorithms with DFL on semantic web are the future work.

Acknowledgments. This research was partially funded by the grants from the Natural Science Foundation of China under grant No.60673092 and No.60873116; the 2008 Jiangsu Key Project of science support and self-innovation under grant No.BE2008044; the Higher Education Graduate Research Innovation Program of Jiangsu Province in 2008 under grant No.CX08B-099Z.and the Project of Jiangsu Key Laboratory of Computer Information Processing Technology under grant No.KJS0820.

References

1. Tho, Q.T., Hui, S.C., et al.: Automatic Fuzzy Ontology Generation for Semantic Web. IEEE Transactions on Knowledge and Data Engineering 18(6) (2006)
2. OWL Web Ontology Language Overview (2004),
   http://www.w3.org/TR/owl-features/
3. Berners-Lee, T., Hendler, J., Lassila, O.: The Semantic Web. Scientific American 284(5), 34–43 (2001)
4. Baader, F., Calvanese, D., McGuinness, D., Nardi, D., Patel-Schneider, P.F. (eds.): The Description Logic Handbook: Theory, Implementation, and Applications. Cambridge University Press, Cambridge (2003)
5. AnHai, D., Jayant, M., Pedro, D., et al.: Learning to map between ontologies on the Semantic Web. In: Proceedings of the 11th International Conference on World Wide Web, pp. 662–673. ACM Press, Hawaii (2002)
6. Calegari, S., Ciucci, D.: Fuzzy ontology, fuzzy description logics and fuzzy-OWL. In: Masulli, F., Mitra, S., Pasi, G. (eds.) WILF 2007. LNCS (LNAI), vol. 4578, pp. 118–126. Springer, Heidelberg (2007)

7. Straccia, U.: A Fuzzy Description Logic for the Semantic Web. In: Sanchez, E. (ed.) Fuzzy logic and the Semantic Web, pp. 73–90. Elsevier, Amsterdam (2006)
8. Li, F.: Research on a Dynamic Fuzzy Data Model. Computer Research and Development 35(8), 714–718 (1998) (in Chinese)
9. Wei, F., Xuefeng, X., Pengpeng, Z., ZhiMing, C.: A Dynamic Fuzzy Description Logic. Wuhan University Journal of Natural Sciences Springer 13(4), 417–420 (2008)
10. Zadeh, L.A.: Fuzzy Logic and Approximate Reasoning. Synthese 30, 407–428 (1975)
11. Sanchez, E., Yamanoi, T.: Fuzzy ontologies for the semantic web. In: Larsen, H.L., Pasi, G., Ortiz-Arroyo, D., Andreasen, T., Christiansen, H. (eds.) FQAS 2006. LNCS (LNAI), vol. 4027, pp. 691–699. Springer, Heidelberg (2006)
12. Stoilos, G., Simou, N., et al.: Uncertainty and the Semantic Web. IEEE Intelligent Systems, 84–87 (September/October 2006)
13. W3C,Web Ontology Language Overview (2006),
    http://www.w3.org/TRowl-features/
14. OntoWeb develiverable 1.3 (2006), http://www.ontoweb.org/
15. Hobbs, J.R., Pan, F.: An Ontology of Time for the Semantic Web. ACM Transactions on Asian Language Processing (TALIP): Special issue on Temporal Information Processing 3(1), 66C–85C (2004)
16. Dazhou, K., Wen, X.B., jianjiang, L., Yanhui, L.: Reasoning for A Fuzzy Description Logic with Comparison Expressions. In: Proceedings of the 2006 International Workshop on Description Logics DL 2006 (2006)
17. Lassila, O., McGuinness, D.L.: The Role of Frame-Based Representation on the Semantic Web, nowledge Systems Laboratory Report KSL-01-02. Stanford University (2001)
18. Stiloe, G., Stamous, G., et al.: A fuzzy Description logic for multimedia knowledge representation (2006)
19. Calegari, S., Loregian, M.: Using dynamic fuzzy ontologies to understand creative environments. In: Larsen, H.L., Pasi, G., Ortiz-Arroyo, D., Andreasen, T., Christiansen, H. (eds.) FQAS 2006. LNCS (LNAI), vol. 4027, pp. 404–415. Springer, Heidelberg (2006)
20. Zhao, X., Li, F.: The Frame of DFL Programming Language. Journal of Communication and Computer 3(1) (2006), ISSN 1548-7709

ADEB: A Dynamic Alert Degree Evaluation Model for Blogosphere

Yu Weng[1], Changjun Hu[1], and Xiaoming Zhang[1,2]

[1] School of Information Engineering, University of Science and Technology
Beijing, 100083, China
[2] School of Information Science and Engineering, Hebei University of Science
and Technology, Shijiazhuang, 050054, China
mr.wengyu@gmail.com

Abstract. This paper proposes a dynamic alert degree evaluation model (named ADEB) for blogosphere to improve the poor tracking ability of topic spreading in traditional topic detection models. The model comprehensively considers the mutual influence of the alert text and the spreading speed, and dynamically modifies the evaluation results by analyzing the alert topic occurring frequency, reviews, comments and existing time of the blog topic. Through tracking the alert blog topic generation and spreading, ADEB effectively avoids the inaccuracy of the static alert limit threshold, shortens the alert response time and improves the detection ability of the burst alert. To validate the performance of ADEB, the experiments on the data corpus about "Campus Network Culture" demonstrate that ADEB has higher application validity and practicality of alert evaluation for blogosphere.

1 Introduction

With rapid development of Web 2.0, blog has become a virtual society format- "Blogosphere". The alert monitoring of blogosphere can effectively collect the public opinion of network, and give the powerful guarantee for keeping the society going on the track of sound progress.

Blogosphere is a collective community of lots of blogs and the related page links. Each blog is made up of a series of topics which is ordered by the published time. The author of blog is named as "blogger" who owns the unique blog sphere. Every blogger could publish the topic logs or comment the others [1, 2]. Some traditional topic detection and tracking (TDT) methods [3, 4, 5] usually emphasis on the web text content analysis, and adopt the certain static empirical weight to calculate the topic alert degree. It could not meet the new demands for the blogosphere alert degree evaluation that the alert generation and spreading not only depend on the web text content, but also be dynamically impacted by the reviews, comments and existing time of the blog topic.

According to the reasons above, a dynamic alert degree evaluation model (named ADEB) for blogosphere is proposed. The model not merely focuses on the alert blog topic text, but also considers the alert spreading by analyzing the reviews, comments and existing time. Through analyzing the alert topic occurring frequency and the

L. Chen et al. (Eds.): APWeb and WAIM 2009, LNCS 5731, pp. 77–87, 2009.

spreading influence, ADEB dynamically modifies the evaluation results and evaluates the blog alert comprehensively. This method effectively avoids the inaccuracy of static alert limit threshold, shortens the alert response time and guarantees the rationality of the alert evaluation in virtual society.

This paper is organized as follows. Section 2 outlines and analyzes the previous approaches of topic detection. In section 3, some problems and the general process are described. In section 4, each part of ADEB is presented in details. In section 5, the experimental results of ADEB are given. Finally section 6 concludes the work with some possible extensions.

2   Related Work

Recent years, topic detection and tracking, as one of the most important technologies for the alert monitoring, has attracted more and more attention. R.Kathleen and McKeow, et al. [6] proposed a novel News Browsing System (named NewsBlaster) in 2002. In 2003, J. Makkonen [7] presented a TDT technology with Spatial-temporal Evidence. With the feature analysis of different topics, the system clustered the same kind of events and realized quick detection of web topics. In 2005, the project of "NewsInEssence" was started in Michigan University. Through the integration of document clustering, topic tracking and multi-documents abstract extraction technologies, the project retrieved the English International News of the World Wide Web [8]. In 2004, by analyzing news data on the Internet, Yu Manqua [9], et al. proposed an effective method of topic detection focusing on the features of events, and an arithmetic named MLCS is offered to organize topics into hierarchical structures. In 2006, JIA Zi-Ya et al. [10] presented an algorithm for news event detection and tracking based on a dynamic evolution model, which adopted the idea of single-pass clustering and combined the specialties of news.

These technologies listed above could effectively improve the manual intervention and strengthen the news topic detection and tracking ability. However, since focusing on the news text itself and neglecting the alert spreading degree evaluation, they could not been applied in the blogosphere alert degree evaluation. The blogosphere alert generation is a dynamic process and is very difficult to evaluate it by a static alert standardization. Differing from the existing work, we present a dynamic alert degree evaluation model which not only calculates the alert degree of blog topic text, but also considers the topic influence degree. According to the occurring frequency, reviews, comments and existing time of the blog topic, the model dynamically modifies the alert degree and more comprehensively evaluates the blogosphere alert.

3   Problems Description and General Process

The generation of blogosphere alert is a dynamic development and change process. ADEB considers the mutual influence of topic text and the alert spreading degree, dynamically modifies the analysis results, real-timely tracks the alert trends and comprehensively evaluates the blogosphere alert. The overview of our approach is illustrated in Figure 1.

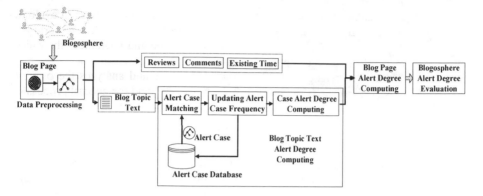

Fig. 1. The general process of ADEB

- **Blog Topic Text Alert Degree Computing**

Blog topic text is the basis of alert generation and spreading. In the blog topic text alert evaluation process, ADEB extracts the topic features of the alert texts and represents them as the alert cases. With the statistical analysis of alert case occurring frequency between target text and alert cases, ADEB dynamically modifies the blog topic text alert degree. This method effectively simulates the alert generation process of the blog topic and has nice practicability.

- **Blog Page Alert Degree Computing**

Considering the mutual influence of the topic text and the blog alert spreading, ADEB real-timely monitors the blog topic reviews, comments and the existing time, analyzes the spreading speed of alert blog page and evaluates the blog page alert degree comprehensively.

- **Blogosphere Alert Degree Evaluation**

Based on the blog page alert degree computing, ADEB evaluates the blogosphere alert. By means of tracking the spreading trends of alert blog pages, ADEB calculates the alert influence of the alert blog pages and evaluates the whole alert degree of blogosphere.

In the following sections, each of the three steps is described in more detail.

4   Dynamic Alert Degree Evaluation of Blogosphere

In the alert evaluation process, firstly ADEB analyzes the blog page structure and extracts the topic text, reviews, comments and the existing time of each blog topic, then dynamically evaluates the alert degree of the blog topic text, blog page and blogosphere [11, 12].

4.1   Blog Topic Text Alert Degree Computing

Blog topic text is the source of alert spreading [13]. ADEB abstracts the diversities of alert features and represents them as the alert cases. With the actual alert case occurring frequency analysis, ADEB modifies the alert case degree and evaluates the blog topic text dynamically. This method greatly stimulates the blog alert generation and guarantees the authority of blog topic text alert evaluation.

4.1.1 Alert Case Representation

With the excellent knowledge representation ability of resource description frame (RDF) [14, 15], ADEB abstracts the blog alert features and represents them as the alert case. The alert case uses statements described as following 3-tuple.

$$Alert\ Case=<Subject,\ Predicate,\ Object>\tag{1}$$

Subject is the alert case and could be marked by a uniform resource identifier (URI). *Object* represents the specific literals. *Predicate* is the binary relations between *Subject* and *Object*. An example of alert case is shown in Fig.2. The alert Case Subject is represented by URI (Http://www.example.org/caseID). Five Predicates are mapped with the related literals.

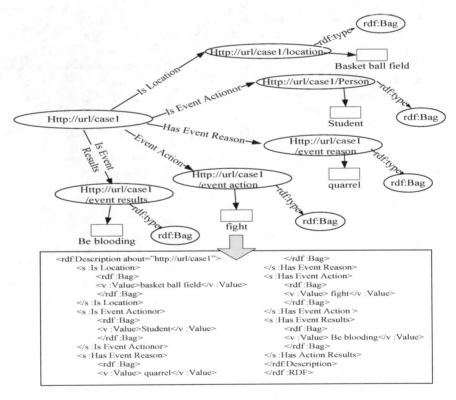

Fig. 2. An example of alert case representation based on RDF

4.1.2 Case Alert Degree Computing

Case alert degree computing is a dynamic modification process which continuously fluctuates by the case occurring frequency. To ensure the feasibility of the iterative process, ADEB sets the initial alert degree for each case in advanced.

Supposed B is a blog page. B_{Topic} is a blog topic of B. A is an alert case. In formula 2, the blog topic text alert degree is calculated by the case alert degree.

$$AlertDegree(B_{Topic}, t+1) = f(CaseAlertDegree(A,t), Match(A,t,t+1)) \qquad (2)$$

Where, t and $t+1$ represent two testing time respectively. $Match(A,t,t+1)$ is the occurring frequency of A within t and $t+1$. Then the case alert case is continuously fluctuated by the case occurring condition. In formula 3, ADEB adopts the exponential function to calculate the fluctuation influence of case occurring frequency.

$$Match(A,t,t+1) = e^{-\left|\frac{Match(A,t+1) - Match(A,t)}{Match(A,t)}\right|} \qquad (3)$$

$P(A, \Delta t)$ represents the variation of the case matching within the interval Δt. With the case occurring frequency analysis, the case alert degree fluctuation could be discussed as follow:

(1) With the increment of matching cases, namely $P(A, \Delta t) > 0$, the case alert degree is increased.

$$CaseAlertDegree(A,t+1) = Match(A,t,t+1) \bullet [1 - CaseAlertDegree(A,t)] + CaseAlertDegree(A,t) \quad (4)$$

(2) With the reduction of matching cases, namely $P(A, \Delta t) < 0$, the case alert degree is decreased.

$$CaseAlertDegree(A,t+1) = Match(A,t,t+1) \bullet CaseAlertDegree(A,t) \qquad (5)$$

(3) If the amount of matching cases keeps invariant, namely $P(A, \Delta t) = 0$, the case alert degree is equal to the one of last time.

$$CaseAlertDegree(A,t+1) = CaseAlertDegree(A,t) \qquad (6)$$

4.2 Blog Page Alert Degree Computing

As a part of blogosphere-an open information communication platform, blog page alert degree computing should consider the mutual influence of alert topic text and the spreading speed. As shown in formula 7, by means of analyzing the increment ratio of the comments and reviews, ADEB evaluates the topic spreading speed of the blog page.

$$TopicSpreadingSpeed(B,t,t+1) = \frac{\sum_{i=1}^{\|B_{Topic}\|} \|Comments_i^{t+1} - Comments_i^{t}\| \bullet \|ReView_i^{t+1} - ReView_i^{t}\|}{\Delta t} \qquad (7)$$

Where, B is a blog page. $\|B_{Topic}\|$ is the amount of topics in B. $Comments\ ^t_i$ and $Reviews\ ^t_i$ represent the amount of comments and reviews in the i-th topic at the time t respectively.

In formula 8, ADEB further integrates the impacts of the blog topic text and the spreading speed and calculates the alert influence of the blog page. Here, ADEB adopts the exponential function to evaluate the alert degree and mappings the fluctuation scope within the range from 0 to 1.

$$AlertDegree(B,t+1) = 1 - e^{-\sum_{i=1}^{\|B_{Topic}\|} AlertDegree(B_{Topic}^i, t+1) \bullet TopicSpreadingSpeed(B,t,t+1)} \qquad (8)$$

Compared with the alert threshold *Th*, ADEB makes the qualitative analysis for each blog pages and calculates the ratio of the alert blogs in the blogosphere.

$$AlertDegree(B,t+1) = \begin{cases} Alert & (greater\ than\ Th) \\ Normal & (less\ than\ Th) \end{cases} \qquad (9)$$

4.3 Blogosphere Alert Degree Evaluation

The blogosphere alert evaluation is based on the alert computing of each blog. According to the actual alert spreading scope statistics, ADEB dynamically calculates the blogosphere alert degree.

Supposed *BSphere* is a blogosphere. $t, t+1$ are the testing time respectively. In formula 10, ADEB analyzes the alert blogs ratio of *BSphere*, tracks the spreading trends of alert blogs and evaluates the alert spreading degree of blogosphere real-timely.

$$AlertSpreading(BSphere,t,t+1) = AlertSpreading(BSphere,t) \bullet (1-e^{\frac{\Delta p}{\Delta(n-p)}}) \qquad (p \le n) \quad (10)$$

Where, n is the whole amount of blogs in the blogosphere. Δp is the increment of the alert blogs. The blogosphere alert spreading degree evaluation could dynamically reflect the current alert spreading of the target blogosphere and give the data support for the alert degree computing.

Finally, ADEB evaluates the blogosphere alert degree. As shown in formula 11, the blogosphere alert degree is evaluated by the average degree of alert blogs and the alert spreading degree.

$$AlertDegree(BSphere,t+1) = AlertSpreading(BSphere,t+1) \bullet \frac{\sum_{i=1}^{p} AlertDegree(B_{alert}^{i})}{p} \quad (11)$$

Here, p is the amount of the alert blogs, B_{alert}^{i} is the i-th alert blog of *BSphere*.

5 Experiments and Analysis

5.1 Experimental Corpus and Prototype

To validate the performance of ADEB, we collected the 80,000 blogs from the blogosphere (www.xiaonei.com) during the time of May 1 to August 30 (named "CNCC", Campus Network Culture Corpus). Table 1 presents the information of CNCC in details.

Table 1. Information of Campus Network Culture Corpus

Time	Total Blogs	Total Topics	Avg Reviews	Avg Comments
May	20,000	187,769	261	16
June	40,000	216,975	294	49
July	60,000	243,681	337	77
August	80,000	343,337	353	92

With the help of alert cases, ADEB evaluates the alert degree of blog topic text. Table 2 shows the alert case types (Alert Speeches, Unhealthy Psychology, Bad Study Styles, Campus Eroticism, Campus Violence and Unhealthy Life Habits) and the amount of each kind.

Table 2. Alert Cases Statistics

Case Type	Num	Case Type	Num
Alert Speeches	311	Campus Eroticism	154
Unhealthy Psychology	282	Campus Violence	243
Bad Study Styles	184	Unhealthy Life Habits	236

Based on that, we developed a prototype to validate ADEB. In Fig.3, ADEB analyzed the topic text, reviews and comments of each blog in CNCC, calculated the alert degree of blog topic and evaluated them according to five alert levels.

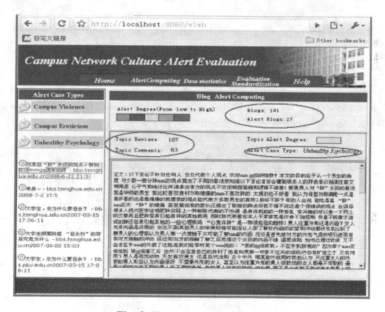

Fig. 3. The prototype of ADEB

5.2 Experimental Results

Testing 1: The Validity Verification of ADEB

As shown in Table 3, we chose "Unhealthy Psychology", "Campus Violence" and "Alert Speeches" as the target cases to evaluate the blog topic alert degree at the time(T1,T2,T3 and T4) respectively. By means of analyzing the topic occurring frequency, average reviews and comments, ADEB tracks the alert fluctuation, dynamically modifies the alert degree and verifies the validity of ADEB.

Table 3. The Validity Verification Testing Data

Case Type	Testing Data	T1	T2	T3	T4
Alert Speeches	Topic Frequency	4,562	2,155	3,521	2,864
	Avg(Reviews)	265	302	353	374
	Avg(Comments)	32	43	71	101
	Alert Level	4.5	2.75	4	3
Campus Violence	Topic Frequency	1,255	3,077	2,268	1,009
	Avg(Reviews)	254	364	378	397
	Avg(Comments)	11	51	75	90
	Alert Level	2	3.5	2.75	1.75
Unhealthy Psychology	Topic Frequency	412	985	1,637	4,461
	Avg(Reviews)	198	235	270	324
	Avg(Comments)	22	55	73	105
	Alert Level	1.4	1.6	2.3	4.7

From the results of the validity verification testing, we noticed that, ADEB could comprehensively consider the mutual impacts of alert text and the spreading speed. In Fig.4, The evaluation results of three alert topics in CNCC are shown respectively. ADEB tracks the process of alert generation and dynamically modifies the alert degree by analyzing the increment ratio of topic occurring frequency, comments and reviews. The method effectively monitoring the alert fluctuation of blog topic and improves the practicability of alert detection.

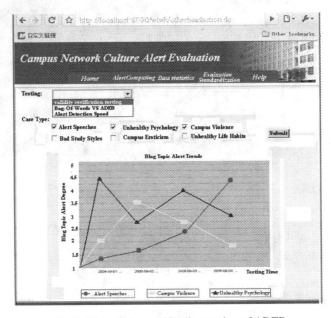

Fig. 4. The validity verification testing of ADEB

Testing 2: The Alert Detection Speed Verification

With the help of CNCC, we made the comparison testing between Bag-Of-Words and ADEB [16]. Analyzing the evaluation results at the time (T1, T2, T3 and T4), the fast alert detection ability will be validated. In Table 4, the inputs data of comparison testing is shown as follow:

Table 4. The Alert Detection Speed Verification Testing Data

Testing Time	Topic Occurring Frequency	Average Reviews	Average Comments
T1	7,622	251	17
T2	11,577	382	45
T3	8,561	243	69
T4	10,343	354	94

In Fig.5, the blog topic alert trend of Bag-Of-Words and ADEB is figured out respectively. From the results of the comparison testing, we notice that ADEB has the higher alert sensitivity and detection speed than Bag-Of-Words.

The blog topic alert trend and the alert blog pages statistics are shown in Fig.5 (A) and Fig.5 (B). At the initial time T1, the alert spreading speed is slow, so Bag-Of-Words and ADEB have the similar alert detection ability. However, with the alert topics continuously spreading (such as T2, T3 and T4), ADEB tracks and evaluates the spreading influence of the alert blog topics. The alert sensitivity is improved greatly. Since only focusing on the alert text analysis and neglecting the alert spreading influence evaluation, Bag-Of-Words could not flexibly fluctuate according with the actual alert. Comprehensively analyzing the mutual influence of the alert text and the spreading speed, ADEB sharply shortens the alert response time and strengthens the detection ability of the burst alert.

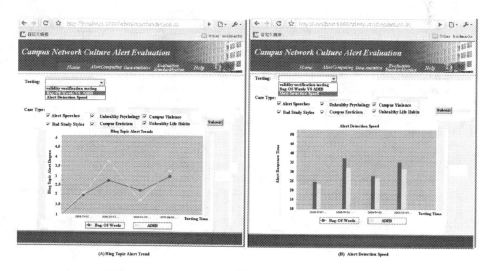

Fig. 5. The Comparison Testing of Alert Detection Speed

6 Conclusions and Future Work

In this paper, a dynamic alert degree evaluation model (named ADEB) for blogosphere is presented. The model considers the mutual impacts of topic text and the alert spreading. Through real-time monitoring the reviews, comments and the existing time of the alert blog topics, ADEB tracks the alert generation and spreading , and comprehensively evaluates the alert degree of blogosphere. Different from traditional topic detection models which adopt the static empirical weight to calculate the topic alert degree, ADEB calculates the blog alert degree by analyzing the alert topic occurring frequency and the influence of spreading speed. This method effectively tracks the alert spreading process in virtual society, strengthens the detection ability of the burst alert and improves the practicality of alert evaluation for blogosphere. To validate the performance of ADEB, we constructed a blog corpus (named "CNCC", Campus Network Culture Corpus) and made the validity and detection speed verification respectively. The experimental results showed that ADEB could dynamically evaluate the alert generation and spreading process, strengthen the burst alert sensitivity and have the practicality of alert evaluation for blogosphere.

In the future, we will further focus on the technique of cooperatively alert evaluation among the inter-network domains and improve the ability of alert spreading analysis by scheduling the local alert evaluation results between different blogosphere.

Acknowledgements

We would like to thank Dr. Liyong Zhao and Peng Shi for discussing some issues about this paper. The work reported in this paper was supported by the Key Science-Technology Plan of the National 'Eleventh Five-Year-Plan' of China under Grant No. 2006BAK11B03 and No. 2008AA01Z109, Natural Science Foundation of China under Grant No. 60373008.

References

1. Blog (2008), http://en.wikipedia.org/wiki/Blog
2. Yu-Hang, Y., Tie-Jun, Z., Hao, Y., De-Quan, Z.: Research on Blog. Journal of Software 19(4), 912–924 (2008)
3. Seo, Y.W., Sycara, K.: Text Clustering for Topic Detection, Carnegie Mellon University, the Robotics Institute (2004)
4. Fiscus, J., Doddington, G.: Topic detection and tracking evaluation overview. In: Allan, pp. 17–31 (2002b)
5. Rebedea, T., Trausan-Matu, S.: Autonomous news clustering and classification for an intelligent web portal. In: An, A., Matwin, S., Raś, Z.W., Ślęzak, D. (eds.) Foundations of Intelligent Systems. LNCS, vol. 4994, pp. 477–486. Springer, Heidelberg (2008)
6. Kathleen, R.M., Barzilay, R.B.: Tracking and Summarizing News on a Daily Basis with Columbia's Newblaster. In: Proceedings of the Human Language Technology Conference, Vancouver, pp. 162–168 (2002)

7. Makkonen, J., Ahonen-Myka, H., Salmenkivi, M.: Topic Detection and Tracking with Spatio-temporal Evidence. In: Proceedings of the 25th European Conference on Information Retrieval Research, pp. 251–265 (2003)
8. Radev, D., Otterbacher, J., Winkel, A.: Newsinessence: Summarizing ONine News Topics. Communications of the ACM 48(2), 95–98 (2005)
9. Manquan, Y., Weihua, L., et al.: Research on Hierarchical Topic Detection in Topic Detection and Tracking. Journal of Computer Research and Development 43(3), 489–495 (2006)
10. Zi-Yan, J., Qing, H., Hai-Jun, Z., et al.: A News Event Detection and Tracking Algorithm Based on Dynamic Evolution Model. Journal of Computer Research and Development 41(7), 1273–1280 (2004)
11. Lam, M.I., Gong, Z., Muyeba, M.K.: A method for web information extraction. In: Zhang, Y., Yu, G., Bertino, E., Xu, G. (eds.) APWeb 2008. LNCS, vol. 4976, pp. 383–394. Springer, Heidelberg (2008)
12. Liu, W., Shen, D., Nie, T.: An effective method supporting data extraction and schema recognition on deep web. In: Zhang, Y., Yu, G., Bertino, E., Xu, G. (eds.) APWeb 2008. LNCS, vol. 4976, pp. 419–431. Springer, Heidelberg (2008)
13. Mei, Q., Cai, D., Zhang, D., Zhai, C.: Topic Modeling with Network Regularization. In: WWW 2008, Beijing, China, April 21–25 (2008)
14. Boley, H., Kifer, M. (eds.): RIF Core Design (2007),
    http://www.w3.org/TR/rif-core/
15. Xiang, M., Junliang, C., Xiangwu, M., Meng, X.: An information retrieval method based on knowledge reasoning. In: Chang, K.C.-C., Wang, W., Chen, L., Ellis, C.A., Hsu, C.-H., Tsoi, A.C., Wang, H. (eds.) APWeb/WAIM 2007. LNCS, vol. 4537, pp. 328–339. Springer, Heidelberg (2007)
16. Bag of words model
    (2008),http://en.wikipedia.org/wiki/Bag_of_words_model

Data Warehouse Based Approach to the Integration of Semi-structured Data

Houda Ahmad, Shokoh Kermanshahani, Ana Simonet, and Michel Simonet

TIMC-IMAG Laboratory
38700 La Tronche, France
houda.ahmad@imag.fr

Abstract. Semi-structured data play an increasing role in the development of the web through the use of XML. However, the management of semi-structured data poses specific problems because semi-structured data, contrary to classical database, do not rely on a predefined schema. The schema of a document is contained in the document itself and similar documents may be represented by different schemas. Consequently, the techniques and algorithms used for querying or integrating this data are more complex than those used for structured data. In this article we propose the architecture of a Data Warehouse designed for the integration of semi-structured data, so as to make possible searches in data repositories of various origins and structures. This architecture relies on the Osiris system, a DL-based model designed for the representation and management of databases and knowledge bases. In particular, we are interested in the Osiris data model, which gives several points of views on a family of objects. On the other hand, the indexing system of Osiris supports semantic query optimization. We show that the problem of query processing on a XML source is optimized by the objects indexing approach proposed by Osiris.

Keywords: XML, data integration, semi-structured data, views.

1 Introduction

During the last ten years, because of the arrival of the Internet and the Web, the number of related data sources as well as the number of potential users of these sources has increased exponentially. These sources are often heterogeneous. In other words they use different models for the representation of the data, such as the relational model, semi-structured models in collections of web pages, systems of files, etc. Beside this multiplicity of supports and formats, the semantics of the various data sources is heterogeneous, which reflects the diversity of the points of views of the systems designer. As a consequence, the languages used for programming or querying these data sources are different. This raises severe problems to the users who try to combine - or "integrate" - information from various data sources.

In this context, companies have to meet two challenges in order to ensure the quality of their data: The fast availability of the information inter-sources and the discovery of the tendencies from the data stored in the company during time. These challenges have led companies to propose a more global homogeneous and coherent

L. Chen et al. (Eds.): APWeb and WAIM 2009, LNCS 5731, pp. 88–99, 2009.
© Springer-Verlag Berlin Heidelberg 2009

vision of their data, which means to realize the integration of data, In other words, companies have to inter-operate various data sources.

Nowadays, data integration approaches take two main directions. The first one is the Data Warehouse approach [1], where a Data Warehouse contains a selective extraction of the relevant information stored in diverse sources. After the construction of the Data Warehouse, the user can formulate his queries over a single database, the Data Warehouse. The second solution is the mediator approach [2], where the data integration is based on the exploitation of abstracted views describing the contents of the various data sources. The data items are not stored at the mediator level and they are accessible only from the original data sources. The mediation approach is preferable if the modifications in the local sources are frequent, whereas the Data Warehouse (materialized) approach is better when the local sources are frequently modified and the query response time must be fast [3].

In this article we focus more particularly on the integration of XML data. We present a materialized framework for XML data integration. This framework is based on the object-based indexing model of the Osiris system to support the optimization of the query processing over an XML source.

2 Integration of XML Documents

XML plays a growing role in the publication of data on the Web. However, contrary to structured data, which are data with a regular structure which can consequently be easily stored in relational tables, semi-structured data, and XML in particular, do not have a priori defined schema. When used, the model underlying semi-structured data can be seen as a relaxation of the traditional relational model, in which a less rigid and less homogeneous structure of the "attributes" is permitted.

The semi-structured model is very useful to represent different kind of documents: multi-media, hypertextual, scientific data, etc. As a consequence to the popularity of XML an enormous amount of XML data of various origins and thus structured in various ways has become available. Heterogeneity is detrimental to the massive exploitation of these data. Indeed, in the absence of specific tools, querying XML data of various origins requires that the different DTDs/schemas and their underlying semantics be known, which is impossible in practice. Consequently, it is necessary to find techniques and methodologies making it possible to question in a simple and effective manner the enormous amount of XML data.

Several projects have dealt this problem like TSIMMIS [4], MIX [5], Xyleme [6][7], VIMIX[7], XyView[9]. The integration process in these systems is usually based on a view mechanism [10]. This mechanism allows defining an abstract (global) schema which presents a unified view of heterogeneous data sources. There are mainly tow approaches to build the global schema of an integration system [11]: The Global As View approach (GAV) defines the global schema as a collection of views over the sources. On the contrary, with the Local As View approach (LAV) the global schema is built regardless of sources. The sources are defined as views over the global schema.

Each one of these two approaches presents advantages and disadvantages. Querying data sources throughout the global schema is easier with *GAV*, while adding new sources is easier with *LAV*.

Our approach uses *GAV* as the method for schema integration; we define the Osiris global schema as a subset of views over the sources and then we transform each document in the source satisfying a concrete schema into a document satisfying a global schema. We use the indexing system proposed by Osiris to optimize the query processing over a XML source. This indexing system will make possible the automatic determination of the views which a document satisfies. This way, the search space for the document answering a query can be contracted to the space of the views satisfying the query. Moreover, we discuss later how our materialized approach can deal with the problem of modifications in a data source (see section 4.2).

3 OSIRIS

3.1 P-Types and Views

Osiris is a view-based database and knowledge base model where views are similar to concepts defined by logical properties, as in Description Logic approaches [12]. The main concept of the Osiris model is the P-type concept, which supports the specification of viewpoints on a domain [13]. For example, STUDENT and TEACHER can be viewpoints of the P-type PERSON in the university domain, To specify a P-type one first gives its minimal (root) view, then its other views by simple or multiple specialization. When specializing a view new attributes and assertions (logical constraints) may be added. The minimal view is the root of the hierarchy of views of a P-type. Thus, in Osiris a P-type is defined from its views, which are object-preserving [14]. Such a top-down approach is contrary to that of relational systems where views are defined as restrictions of a set of existing relations, and may themselves be used as relations in order to define other views.

The type of a P-type is derived from the views declarations (including the minimal view). The type person contains all the attributes and methods which appear in its views. The domain of an attribute in the type person is the union of its domains in the views where it is declared.

To express that a person may be seen as a student, a teacher, a sportsman, one will create the views PERSON, STUDENT, TEACHER, … as subtypes of the P-type PERSON. The set of interest of the minimal view PERSON is identical to that of the P-type PERSON. The domain of another view is a subset of the domain of the view it specializes, or of the intersection of the domains of the views it specializes in case of multiple specialization.

In OSIRIS, a P-type is given the name of its minimal view. All the objects of a P-type are models of its minimal view. Access to an object under a viewpoint provides access to the attributes of the viewpoint. Thus accessing an object from the minimal view only provides the attributes of the minimal view while accessing it in the viewpoint of the P-type gives access to the whole set of attributes of the type.

An object belongs to one P-type, for example PERSON, if and only if it satisfies the requirements of its minimal view. This means that its assertions are valid. An object can belong to only one P-type, which means that P-types are disjoint concepts if considered in a DL perspective.

We present the main features of the P-type description language through a very simple OSIRIS example. The modeled universe is that of persons and vehicles. Persons may be STUDENT, TEACHER, TRAINEE-TEACHER, etc, or some of them simultaneously. A given person is a model of the minimal view and may belong to none, any or several other views.

```
Type  ADDRESS:  (street:  STRING;  postcode:
          STRING;
                  city: STRING);
View PERSON - - Minimal view of P-type PERSON
attr   Name: P_NAME; - - P_NAME is declared
       elsewhere
   Children: setof PERSON;
   Sex: CHAR;
   Age: INT;
   Ad_pers: ADDRESS;
   Military_Service: STRING;
   IncomeTax:   REAL   calc;--   procedural
       attachement
   CarsOwned: setof CAR; - - CAR is a view of
                      -- a P-type VEHICLE
Key   Name      -- External key; not mandatory
methods
          -- Other functions specification
assertions
-- Domain Constraints
       Sex in {"f", "m"};
       0 ≤ Age ≤120;
       Military_Service in {"yes", "no", "deferred",
       "exempt"};
end; -- Note that the minimal view automatically
       contains a private attribute OID : toid.
view   STUDENT:  PERSON   --   STUDENT
       specializes PERSON
```

```
attr   Studies: STRING in {"graduate", "postgraduate",
       "doctorate"};
       Year : int ;
end ;
view   TEACHER:  PERSON--TEACHER  specializes
       Person
attr   Diplomas : setof STRING  in {"degree",
       "B.A.", "BSc" ,"MA.","MSc","PhD"} ;
       Status: STRING in {"trainee", "lecturer",
       "professor", "instructor", "doctor" };
end;

view   PROFESSOR:   TEACHER--   specializes
       TEACHER
assertions  Diplomas contain "PhD";
            Status = "professor";
end ;

view TRAINEE-TEACHER: STUDENT, TEACHER -
       specializes STUDENT and TEACHER
assertions
       Age ≤ 27;
   Status = "trainee";
   Studies = "graduate";
   Diplomas contain "degree";
End ;
```

3.2 Classification Space

Most innovative features of the system come from the use of a classification space, which is distinct from the original set of users' views.

The classification space is a partitioning of the object space into equivalence classes named Eq-classes, according to the relation "have the same truth values according to the (entire set of) Domain Predicates of the type". As a consequence all objects of a given Eq-class are models of the same assertions (Domain Constraints and Inter-Attribute Dependencies) [13].

Construction

In a P-type t, one considers for each attribute A_i the set $PT(A_i)$ of predicates over A_i which appear in the assertions (Domain Constraints and Inter-Attribute Dependencies) of the views of t. Elementary predicates in these constraints are of the form A_i

$\in D_{ik}$ where D_{ik} is a subset of the domain of definition Δ_i of A_i. A predicate $A_i \in$

D_{ik} defines a partitioning of Δ_i into two (disjoint) subdomains: D_{ik} and Δ_i - D_{ik}. The product of all the partitions [16] defined by the predicates of $PT(A_i)$ constitutes a partition of Δi whose blocks dij, called Stable Sub Domains (SSDs), have the following property:(Stability of an attribute) When the value of an attribute Ai of an object varies within the same stable subdomain dij, Ai continues to satisfy the same set of predicates of $PT(Ai)$.

Considering the above definition of the P-type PERSON, and considering only the predicates on the attributes Age, Military_service and Sex, we obtain the following partitioning of the attributes:

SSDs of Age: $d11 = [0, 18[, d12 = [18, 27], d13 =]27, 65[,$
$d14 = [65,140]$
SSDs of Sex: $d21 = \{"f"\}, d22 = \{"m"\}$
SSDs of Military_ service: $d31 = \{"yes"\},$
$d32 = \{"no", "deferred, "exempt"\}$

This partition can be extended to the space of objects (which is restricted here to the three dimensions considered) and constitutes the classification space of the P-type.

Each element of the classification space is called an Eq-class. It is represented by a tuple with n elements, where n is the number of classifying attributes of the P-type.

Let $SSD_{Attr1}, SSD_{Attr2}, ..., SSD_{AttrN}$ be the set of the stable subdomains of the attributes $ATTR1, ATTR2, ..., ATTRn$ of the P-type T respectively.

ClassificationSpace $T \subseteq \{<d1i, d2j, ..., dnk> |$
$d1i \in SSD_{Attr1}$ and $d2j \in SSD_{Attr2}$ and $...$
$dnk \in SSD_{AttrN}\}$

Partitioning the object space into Eq-classes is central to the implementation of the Osiris system. Although the actual partition is not represented in its totality (its size is exponential to the number of classifying attributes) it underlies most runtime processes such as object classification, view classification (subsumption), integrity checking and object indexing.

3.3 Indexing Structure Descriptor

An indexing structure called ISD (Indexing Structure Descriptor) is defined for each P-type [13]. Its main components are:

A vector of SSDs representing one or more Eq-classes indexing a set of objects. Two values have been added to represent the unknown and undefined (null) states of an attribute.

A vector of views that provides the status (Valid, Invalid or Potential) of each view of the P-type for the set of objects indexed by this ISD.

A reference to the actual set of objects of the ISD.

The total number of objects indexed by the ISD.

An object, even if only partially known, belongs to one and only one ISD, as an ISD denotes all its possible Eq-classes. The sets of Eq-classes corresponding to different ISDs may not be disjoint in the case of incompletely known objects. Only actual ISDs, i.e., ISDs containing actual objects, are represented. When an unknown attribute becomes known the corresponding object changes its ISD (a new ISD is created if necessary). When an attribute changes its value it remains within the same ISD iff none of its attributes has changed its SSD.

3.4   Query Evaluation in Osiris

Queries are evaluated in three steps:

1. Determination of the ISDs corresponding to the query when rewritten in terms of the SSDs of its attributes.
2. Determination of the ISDs indexing objects that are valid for the query.
3. Projection of the resulting objects onto the attributes of interest of the query.

4   Materialized Approach for XML Data Integration

The system architecture proposed by our approach for the integration of XML data is shown in Figure 1. Another study is made to integrate structured data sources with a hybrid approach basing on the Osiris indexing system [3]. The system consists of:

1. A transformation processor, which transforms an Osiris schema into a XML schema.
2. A wrapper for every concrete XML schema, which contains the mapping between this concrete schema and the global Osiris schema. This mapping between schemas is a path-to-path mapping.
3. A transformation processor, which transforms each XML document in the data source satisfying a concrete schema into a document satisfying the global schema.

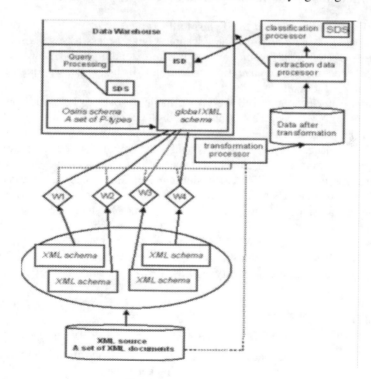

Fig. 1. System architecture

4. An extraction processor, which parses each XML document obtained after transformation in order to extract atomic values.
5. A classification processor, which classifies objects after their extraction.
6. A Data Warehouse, which stores the data corresponding to the global schema and extracted from the local sources.
7. A query processing, which uses the information of Stable Sub Domains (SSDs) to obtain the ISDs satisfying the query.
8. An extraction processor which searches objects satisfying the query and extracts the relevant data for the query.

The Data Warehouse stores data in relational tables; it contains a Query Processing and the following three memory spaces:

Global Osiris schema: it contains the integration schema, which is defined as an OSIRIS schema. It consists of one or several P-types; each P-type is a hierarchy of views with its proper constraints.

SSD space: it is used to save the Stable Sub-Domains of the system. It is used by the query processing to obtain the ISDs which satisfy a query.

ISD space: it is used to save the object indexing information.
In the following section, we explain in detail every phase in this integration system.

4.1 The Different Steps of the Integration Process

4.1.1 XML Representation of the Type of a P-Type

In its initial version, the global mediator schema giving access to several source data bases is an Osiris schema. For this reason, before extending the system to take into account XML documents, we have tried to determine if this model was adapted to the integration of XML documents.

The concept of view, central to the P-type model, proves to be a key concept for the integration of XML documents. Indeed, with the concept of view, it is not necessary that a DTD/schema matches the whole P-type, but only a subset of its views; the minimal view is necessarily a part of this subset. An immediate consequence is that very different DTDs can be in correspondence with the same P-type. Consequently, the use of the P-type concept allows an easier management of heterogeneous DTDs.

Since the Osiris system is based on the P-type notion, the first question is: is it possible to translate a P-type in Osiris into an XML schema\DtD and vice versa?

To answer this question we have considered the type of the P-type (Person) of our example and we have used the W3C XML schema recommendation to represent it in XML. The conclusion of our study is that we can represent the type of a P-type in XML. We show in the following the type of the P-type (Person) and its corresponding schema.

Type of the P-Type PERSON
The XML schema in Fig.2 represents the type of the P-type (PERSON):

As shown in Fig. 2, the standard XML schema gives the opportunity to represent the different types of data: simple types (string, integer, float, etc.) and complex types (e.g., TypeAddress: street, postcode, and city). The XML schema also gives the

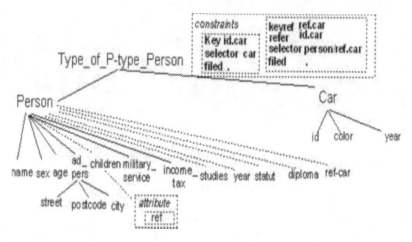

Fig. 2. XML Schema of the type of PERSON

possibility to determine the frequency of an element in the XML document corresponding to schema (see for example Fig.3). The minimum view's elements in our example (name, sex, age, ad_pers) are mandatory (not null) elements in the document; they must appear one time and we represent them with bold lines. The elements represented by pointed lines are optional, they can appear 0,1 or several times in the document; for example, the elements (studies, year, status, diploma) are optional.

The XML schema offers the possibility that an element refers to another element. For example the attribute ref, in the element children, whose type is IDREFS, refers to the attribute ident, in the element Person, whose type is ID.

Moreover, the XML schema offers the possibility to use the identity constraints (key and keyref) similar to the concepts of primary and foreign key in the relational model. In our example, the element ref-car in the element person can refer to one or several elements id in the element car.

The example represented in this section showed us the possibility to represent the global Osiris's schema by using a XML schema. In the following section we pass to the following step of our integration processor based on Osiris. The role of this phase is to establish the correspondence between the global schema and the concrete XML schemas in the source base.

4.1.2 Correspondence Between the Global Osiris Schema and a Concrete XML Schema

We suppose that the source database contains XML documents which represent data of a specific domain. Every group of documents satisfies a XML schema, which we call a concrete XML schema; for the data source we have a set of concrete XML schemas which have different structures.

In this step we put the global Osiris schema (the schema obtained after the last step) and the concrete XML schemas in the source base. Then we establish the correspondence between the global Osiris schema and every concrete schema representing a group of XML documents in the source base.

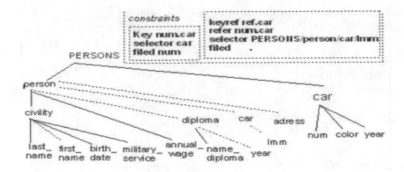

Fig. 3. The concrete XML Schema

Let us consider the Osiris schema in our example (Fig.2) and suppose that we have the concrete schema represented in Fig.3 in the source base.

We use the path-to-path approach to establish the correspondence between the concrete schema and the abstract global schema, because is proved to be better than the other approaches (DTD-to-DTD and tag-to-tag) [15].

We describe in the table below the correspondence examples between the Osiris attributes and the elements of concrete XML schema of our example.

Table 1. Correspondence between the Osiris attributes and the elements in the XML schema

P-type, Osiris Attribute	XML Schema, Element
1. PERSON, name	PERSONS\Person\Civility\last_name
2. PERSON, age	PERSONS\person\civility\birth_date
3. PERSON, diploma	PERSONS\person\diploma\name_diploma
4. PERSON, ref-car	PERSONS\person\car\Imm
5. PERSON, street	PERSONS\person\adress
PERSON,code_post	
PERSON, city	

4.1.3 Transformation Processor

To be able to take advantage of the indexation offered by Osiris, The role of this phase is to transform every concrete XML document satisfying a concrete schema in a document which satisfies the Osiris global schema. This transformation is made in a semi-automatic manner by using the correspondence information generated during the previous phase.

We suppose that we have m groups of XML documents: group1, group2, ..., groupm satisfying the concretes schemas schemac1, schemac2,, ..., schemacm respectively, we have the global schema schemag and the mappings: map1, between schemac1 and schemag, map2, between schemac2 and schemag,..., and mapm, between schemacm and schemag.

Each XML document in the source base belongs to a group $group_i$ where $1 \leq i \leq m$ and satisfies the concrete schema $schma_{ci}$. The transformation processor transforms this document into a document satisfying the schema schma$_g$ by using the mapping information map$_i$ where $1 \leq i \leq m$.

4.1.4 Extraction Processor

Following to the previous phase, all the documents of the source base are transformed into documents satisfying the Osiris global schema. During the extraction phase we parse these documents in order to extract the OSIRIS objects and we store these objects at the Data Warehouse level. We also store the Oids of these objects and the identifier of the corresponding element in the base source and the identifier of the document which contains this element in the base source. We store this information in a relational table; the schema of this table has the following form:

- Data _Store (oid, , elemID docID, attr1, attr2, …, attrm):
- Oid is the object identifier in the Osiris's system.
- elemID is the identifier of the corresponding element in the source base
- docID is the identifier of the document which contains the corresponding element in the source base.
- attr1, attr2, …, attrm are the classifying Osiris attributes and all the simple elements having atomic values in the source base.

We will use this information to retrieve the query answer and also in the case of data modification in the source (see section 4.2).

4.1.5 Classification Processor

The Osiris system performs object classification as explained in section 2. The classification processor uses SSD information and classification results are sent to the ISD (Indexing Structure Descriptor). We have to repeat this process at each schema global evaluation (in case of schema modifications or assertions changes).

4.1.6 Query Processing

The user poses his query over the Osiris global schema. The system uses SSD information in order to evaluate the query as explained in section 3 by performing the following steps:

1. the system rewrites the query (in terms of) by SSDs and determines the potential ISDs corresponding to the query.
2. among these ISDs, the system determines the ISDs which contain the objects validating the query (valid ISDs) and those which contain the objects that could potentially participate in the response of the query (potential ISDs). In these cases there are other conditions to verify. If an object satisfies these conditions, it participates to the response.

After these two steps, we determine the effective ISDs of the query, i.e., those stored in the ISD space of the system because they have at least a stored object. The valid ISDs give the valid objects for the query, whereas the potential ISDs give the objects for which a supplementary test must be made to determine if it is an actual answer to the query. For the potential objects we define the conditions to verify, so that an object of the ISD satisfies the query.

In the ISD (Indexing Structure Descriptor) of our system, for every object we have a reference towards the corresponding object stored in the Data Warehouse. For the objects validating the query we use their reference to be able to recover the atomic values of attributes asked in the query. For the possibly valid objects, we need to verify that these objects satisfy the supplementary conditions that are necessary to confirm whether or not the object satisfies the query.

4.2 Source Data Modification

In the case of modification of a data in a document d in the source, we propose the following algorithm:

We suppose that d' is the document after modification, let T be the table where we stored the data as explained in (section 4.1.4). We repeat the transformation and the extraction processor for the document d', we obtain the table T', then we compare the two tables T and T'. For every tuple in the table T containing the identifier of the document d:

1. if there is an element deleted in the table T', we delete the corresponding tuple in the table T and its corresponding oid in the ISD.
2. If there is a modification of the value of a classifying attribute, we modify the values in the corresponding tuple in the table T, then we activate again the classification processor to classify the modifying object in the ISD.
3. If there is a simple modification of a non classifying attribute we change only the value of the attribute in T.
4. If there is a new object, we add the corresponding tuple in the table T and we classify the corresponding object in the ISD.

5 Conclusion

In this article we have proposed a method to integrate XML documents into a semantic middleware. The key step of this approach is the definition of correspondences between a global Osiris schema and several concrete XML schemas. To do so, we use an intermediate step in order to 1) define the types of a P-type, i.e., its maximal type and the types of each of its views, and 2) translate the resulting type into XML schemas.

The schemas issued from a P-type are not independent but are linked by hierarchical relationships. Hence, for each concrete schema or DTD, it is possible to calculate the correspondences between the tags of the schema corresponding to the maximal type and the tags of a concrete schema. We use path-to-path correspondences, as they have proved to provide the best results. On the establishment of a correspondence, a procedure can be attached to perform some calculation when necessary, e.g., calculate age value from birthday date.

We also showed that our approach uses the indexing structure (ISDs) offered by Osiris to perform a semantic optimization of queries. This indexing makes possible the automatic determination of the views that a document satisfies. This way, the search space for the documents answering the query can be contracted to the space of the views satisfying the query. This component is important because the actual exploitation of a large number of documents remains a problem, even if the principle of using an abstract DTD constitutes an important advance for the exploitation of XML documents in a homogeneous manner.

We also showed how our materialized approach maintains the modification of source data.

References

[1] Wu, M.C., Buchmann, A.P.: Research issues in Data Warehousing. In: Datebanksysteme in Buro, Technik and Wissenschaft, pp. 61–82 (1997)

[2] Wiederhold, G.: Mediators in the Architecture of Future Information System. IEEE Computer Magazine 25(3), 38–49 (1992)

[3] Kermanshahani, S.: Semi-Materialized Framework: a Hybrid Approach to Data Integration. In: CSTST Student Workshop, Paris (October 2008)

[4] Garcia-Molina, H.: The TSIMMIS Approach to Mediation: Data Models and Languages. Journal of Intelligent Information Systems 8(2), 117–132 (1997)

[5] Bornhovd, C.: MIX – A Representation Model for the Integration of Web- Based Data. Technical report, Dep.CS, Darmstadt University of Technology, Germany (1998)

[6] Abiteboul, S., Cluet, S., Ferran, G., Rousset, M.C.: The Xyleme Project. Gemo Repot 248, INRIA (2001)

[7] Manolescu, I., Florescu, D., Kossman, D.: Answering XML Queries Over Heterogeneous Data Sources. In: Proceeding of 27th International Conference on VLDB (2001)

[8] Baril, X.: Un modèle de Vues pour l'Intégration de Sources de Données XML: VIMIX. PHD thesis, Languedoc University of Science and Techniques (2003)

[9] Sebi, I.: Interrogation de Documents XML à Travers des Vues. PhD thesis, EDITE, CEDRIS Laboratory (2007)

[10] Cannataro, M., Cluet, S., Tradigo, G., Veltri, P., Vodislav, D.: Using Views to Query XML. In: Encyclopedia of Dtabase Technologies and Applications, pp. 729–735 (2005)

[11] Halevy, A.: Answering Queries Using Views: A Survey. The VLBD Journal 10(4), 270–294 (2001)

[12] Roger, M., Simonet, A., Simonet, M.: Bringing together description logics and database in an object oriented model. In: Hameurlain, A., Cicchetti, R., Traunmüller, R. (eds.) DEXA 2002. LNCS, vol. 2453, p. 504. Springer, Heidelberg (2002)

[13] Simonet, A., Simonet, M.: Classement d'Instance et Evaluation des Requêtes en Osiris. In: BDA 1996: Bases de Données Avancées, Cassis, France, pp. 273–288 (August 1996)

[14] Scholl, M.H., Laasch, C., Tresch, M.: Updatable Views in Object-Oriented Databases. In: Delobel, C., Masunaga, Y., Kifer, M. (eds.) DOOD 1991. LNCS, vol. 566, pp. 187–198. Springer, Heidelberg (1991)

[15] Ahmad, H., Kermanshahani, S., Simonet, A., Simonet, M.: A View-Based Approach to the Integration of Structured and Semi-structured Data. IEEE International Baltic Conference on Databases and Information Systems-Communication of Baltic DBIS (2006)

[16] Stanat, D., McAllister, D.: Discrete Mathematics in Computer Science. Prentice Hall, Englewood Cliffs (1977)

Towards Faceted Search for Named Entity Queries

Sofia Stamou and Lefteris Kozanidis[1]

Computer Engineering and Informatics Department, Patras University, Greece
{stamou,kozanid}@ceid.upatras.gr

Abstract. A considerable fraction of the web queries contain named entities. This, coupled with the fact that a proper name might refer to multiple entities, imposes the ever-increasing need that search engines handle efficiently named entity queries. In this paper, we present a technique that automatically identifies the distinct subject classes to which a named entity query might refer and selects a set of appropriate facets for denoting the query properties within every class. We also suggest a method that examines the distribution of the identified query facets within the contents of the query matching pages and groups search results according to their entity denotation types. Our preliminary study shows that our technique identifies useful facets for representing the named entity query properties in each of their referenced subject classes.

Keywords: faceted search, named entity queries, Wikipedia corpus.

1 Introduction

The key objective of all Information Retrieval systems is to help the users find the desired information about their search requests in an effortless yet successful manner. With the proliferation of the web content, search engines have become an indispensable tool in the information seeking process. Despite the popularity and the tremendous capacity that search engines have nowadays, there are still open issues concerning their ability to satisfy all user needs. This is essentially due to vocabulary mismatches between the indexed documents and the user issued queries that hinder the engines' ability in detecting the underlying correlation between documents and queries.

In this paper, we suggest a method for identifying the referred concepts of Named Entities (NE) in both user queries and query matching pages, in order to improve the engines' ability in handling named entity queries. What motivated our study is on the one hand the observation that a significant portion of popular web queries contain NEs [1] [2] and on the other the fact that different entities might be verbalized in user queries with the same name. As examples, consider the NE queries *Java* which might intend the retrieval of information about the programming language or the coffee, and *Apple* which might intend the retrieval of information about the company, the

[1] The authors is financially supported by the PENED 03ED_413 research grant, co-financed by 25% from the Greek Ministry of Development-General Secretariat of Research and Technology and by75% from the E.U. European Social Fund.

L. Chen et al. (Eds.): APWeb and WAIM 2009, LNCS 5731, pp. 100–112, 2009.

computer or the fruit. As the examples indicate, a NE query might refer to different entity types, but in the absence of any implicit *knowledge* about the different reference classes of named entities, the engine would always retrieve the same results; usually a mixed list of pages about the distinct query denotations. Consequently, users would have to go over a long list of results and access their contents in order to satisfy their search intent.

To overcome such difficulties, researchers have proposed a number of techniques for personalizing search results according to specific user interests. Personalization, although it might work well for some users or queries, nevertheless it entails practical limitations in a real deployment as it pre-requisites processing large volumes of web transaction logs in order to model the user search preferences. Another possible direction towards helping users find the desired information is to classify web pages into faceted hierarchies and present the query matching results grouped together according to the corresponding query senses [7]. Although faceted search was primarily addressed in the context of cataloguing and library systems, nowadays it is widely employed by e-commerce applications (e.g. amazon.com, shopping.com) and recently it has attracted the interest of the web search community. This surge of interest is basically because facets allow a document to exist simultaneously under different categories, each representing distinct document properties, and enable the user access the categorized documents in multiple ways.

Bearing in mind the power and success of existing faceted search approaches, we introduce the use of facets for describing the different entity types to which NE queries might refer and for detecting within the contents of every query matching page the specific subject that query entities denote. To enable that, we start by defining a set of facets that are good candidates for representing the different types of NE queries. In this respect, we rely on the Wikipedia corpus[2] and a number of heuristics for annotating every NE query with one or more suitable facets. Then, we examine the NE query results in order to estimate the distribution of the query facets within the contents of every query matching page. Based on the underlying association between the facets of the queries and their returned pages, we propose grouping search results according to the query entity types that their contents represent. Grouped search results accompanied by their derived faceted terms, when displayed to the users can help them find the information sough in an effective yet efficient manner. Our preliminary study shows that our technique delivers useful facets for representing the subject classes that NE queries might represent. Thus, we claim that our approach could be fruitfully explored towards improving the users' search experience when querying the web for named entities.

The remainder of the paper is organized as follows. We begin our discussion by reviewing related work. In Section 3, we present our approach towards the automatic identification of useful facet terms for representing the NE query classes. In Section 4, we propose a faceted search approach that groups query results according to the representation of the query facets in their contents. In Section 5, we report the results of our experimental evaluation and we conclude the paper in Section 6.

---

[2] http://en.wikipedia.org/wiki/Wikipedia:Database_download

2 Related Work

Many researchers have studied the problem of named entity recognition and disambiguation. To address the problem, researchers have proposed numerous ways for exploring entity-local features [3] [4], or complex lexico-syntactic and morphological data [5] [6] in order to detect NEs within text documents. Recently, [13] [7] suggested the use of encyclopedic knowledge for resolving the different classes to which a NE might refer. In their work, [7] utilize the Wikipedia corpus in order to derive a dictionary of named entities and then based on the detected entities' contextual elements and position in the Wikipedia taxonomy, they determine a set of candidate concepts that every entity denotes. They disambiguate NEs based on the degree of correlation between the entities' contextual elements and their conceptual categories. In a different approach, [8] studied the enrichment of Wikipedia pages with named entities. In this respect, they designed a method that extracts a number of features (both contextual and page-based) from the Wikipedia pages and uses them as training examples for a classification module. Upon training, the classifier assigns every Wikipedia page with an appropriate tag (from a pre-defined list) that is reflective of the named entities contained in the page. In a similar task, [9] employed a graph-based approach and experimented with categorizing named entities in the Japanese version of Wikipedia. Recently, [17] proposed a model for representing web pages as sets of named entities that are generally informative of the following subject types: *person*, *location*, *organization* and *time*. Based on the inter-relations between these subject types, the authors determine which named entities suffice for representing the content of web pages.

Our work also relates to some recent studies on faceted search and facet terms extraction. More specifically, in [10] and [11] the authors introduce a method for identifying useful facets for browsing textual databases. Their method relies on WordNet [12] hypernyms, Google search results and the Wikipedia pages for defining a set of broad terms with which to expand keyword terms extracted from the database contents. In [14] the authors propose the faceted query logs analysis and present a method for classifying web queries into the following faceted categories: *ambiguous*, *authority*, *temporally-* and *spatially-sensitive* requests. In [15] a faceted search personalization approach is discussed, emphasizing on how to customize the search interface according to user ratings. In [16] a data-driven technique, called Dynamic Category Sets (DCS), is introduced that discovers sets of values across multiple facets that best match the query and enables the user disambiguate the latter via a search-by-category clarification dialog. In a recent study, [18] introduce a dynamic faceted search system that automatically discovers a small set of valued facets that are deemed interesting to a user and enable the latter understand the important patterns in the query results.

Although our study touches upon issues that have been previously addressed, it is different from existing works in the following: we introduce a novel facet extraction technique that is specifically tailored for representing the classes of NE queries. Our method uses a small amount of NE contextual data from which it extracts useful facets. To estimate the usefulness of the identified faceted terms, we propose a metric that estimates how valuable a facet is for representing the named entity properties within a particular subject class. We also suggest the exploitation of the identified NE query facets while looking for query relevant data. In this respect, we suggest a method that relies on the distribution of both the query keywords and the query facets within the query matching pages in order to group search results according to their named entity denotations. In the following section, we discuss the details of our facet selection approach in order to represent the subject denotations of NE queries.

3 Faceted Representation of Named Entity Queries

In this section, we discuss the details of our work towards associating named entity queries with a set of useful facets for denoting the properties of their refereed classes. There are two main challenges associated with our goal: how to identify all possible subject classes to which a NE query might refer and how to select useful terms for verbalizing the NE query properties in each of the identified classes. For the first challenge, we explore the Wikipedia corpus and we apply a number of heuristics for determining a set of features that are informative of the NE subject classes (Section 3.1). For the second challenge, we introduce a facet extraction algorithm that explores the entities' contextual elements in order to derive useful terms for describing the NE properties in each of the referenced classes (Section 3.2). Before delving into the details of our method, let's describe the process we adopt for identifying NE queries.

For any query q, we firstly wish to determine whether q is a named entity. To be able to judge that, we rely on the Wikipedia pages in the titles of which we look for the query keywords. Upon their detection, we download the contents of the respective pages and we follow the steps suggested by [7] in order to assess whether a given term corresponds to a NE or not. These steps summarize to (quoting from [7]): (i) *If q is a multiword query, check the capitalization of all content terms in q. If all contents words of q appear always capitalized in their corresponding pages then q is a NE.* (ii) *If q is a single-term query and contains at least two capital letters, then q is a NE.* (iii) *Count the occurrences of q terms in the text of the page, in positions other than at the beginning of the sentences. If at least 75% of these occurrences are capitalized, then q is a NE.*

The combination of the above steps in the work of [7] resulted to the extraction of nearly half a million named entities form the Wikipedia corpus. In our study, we apply the same process in order to automatically identity whether a search request is a named entity query or not. Note that this NE detection heuristic applies to languages that do not capitalize common nouns, e.g. English. Based on the identified NE queries, we designed a method that automatically derives the different subject classes to which a named entity might refer. The details of our method are discussed next.

3.1 Deriving the Named Entity Query Classes

The first step towards discriminating between the different subject classes to which a named entity query might refer is to examine whether the identified NE has a unique type of reference or not. In particular, we need to investigate if a NE always refers to a single class of subjects (e.g. humans, locations, etc.) or not. For our investigation, we start by looking at the presence of clarification sentences in the Wikipedia articles that discuss the given NE. Clarification sentences always appear at the beginning of a Wikipedia article right after the title section and are of three generic types, stating:

a) ***This article is about*** [`clarification term/phrase`]
b) ***For other uses, see*** (`link to a disambiguation page`)
c) ***For other*** [`clarification term/phrase`] ***with the same name, see*** (`link to a disambiguation page`)

In some cases, a clarification sentence might be a combination of the above types; usually of types (a) and (b). Based on clarification sentences, we can distinguish two main groups of NEs: those which are clarified in Wikipedia (i.e. their corresponding articles contain clarification sentences) and those which are not clarified.

For named entities with clarification sentences, we examine the type of their clarifications and we further sub-group them according to the following criterion: entities whose clarification sentences are all of type (c) listed above, are deemed as entities with a single reference class, where the latter is expressed via the clarification term(s). We refer to such entities as **single faceted NE** and we further process them to verbalize their facets as we will present in the following section. Named entities with clarification sentences of types (a), (b) or a combination of the above-listed types are intuitively considered to be expressive of multiple subject classes and we refer to them as **multi-faceted NE** (one clarification per subject class). Note that under our approach a disambiguation sentence[3] is generally deemed as a clarification one, assuming that Wikipedia contributors edited such sentences for clarifying the usage of a given entity. Therefore, a NE is deemed to be referring to as many classes as the number of the sentences in which the entity's name is clarified. In the following section, we present how to select useful faceted terms for denoting the different classes to which a NE might refer.

Now, let's go back to the group of named entities that lack clarification sentences in their corresponding Wikipedia articles. To identify which of these NE are single faceted and which are multi-faceted, we extract the definition sentences from their respective Wikipedia articles. A definition sentence is the first sentence in the body of the article that contains the named entity and a wordform of the verb *to be* followed by one or more common terms and/or disambiguation entities[4]. Relying on the extracted definition sentences, we compute their feature vectors within a sliding window of ten words surrounding the entity reference. For our computations we employ a bi-gram model that records for every word (i.e. feature) in a definition sentence its relative position around the entity and the number of times the word appears in that position. Then, based on the sentences' overlapping features in the derived vectors, we organize them into clusters of shared elements. For example, a significant portion of the Wikipedia definition sentences about humans, exhibit the feature (`born`) right after the entity's name. Under our criterion, all definition sentences containing the above feature are grouped together. Following the above steps, we organize every Wikipedia definition sentence about a NE into one or more clusters, depending on the number of their shared features.

Based on the number of clusters to which a named entity's definition sentence is assigned, we make the following assumption: if the definition sentence of a NE belongs to a single cluster, then the NE is single faceted. For example, a NE whose

---

[3] A disambiguation page in Wikipedia is a page that contains the term "disambiguation" in its title and contains several possible disambiguations of a term [7].

[4] According to [8], a disambiguation Wikipedia entity is a hyperlinked term that points to another Wikipedia page in which the meaning of the term is resolved.

definition sentence is organized under a single cluster, which groups sentences sharing the feature $t = holiday$ (e.g. Christmas is an annual *holiday*) is deemed as expressive of a single class of subjects. Conversely, if the definition sentence of a NE belongs to multiple clusters, then the NE is multi-faceted and refers to as many classes as the number of clusters to which the definition sentence of the NE has been assigned. For example, a NE whose definition sentence is grouped into (say) two clusters, one of which groups sentences that share feature $t_i = comic\ book$ (e.g. *Superman is a fictional comic book superhero .... cultural icon*) and the other groups sentences that share feature $t_j = cultural\ icon$ (e.g. *Superman is a fictional comic book superhero widely considered to be one of the most recognized of such characters and an American cultural icon*) is deemed as being expressive of two object classes, e.g. book and cultural icon. Based on the process described above, we can discriminate between named entities that always refer to a single class of subjects and those that might represent different subject classes in their contextual denotations. Following on from that, we need to define a set faceted terms that are useful for representing the NE properties in each of the identified subject classes. In this respect, we have designed a facet selection algorithm; the details of which are discussed next.

3.2 Selecting Faceted Terms to Represent the Named Entity Query Classes

In selecting a set of useful facets for describing the different types of named entity references, we rely on the contextual elements in the NEs' clarification and definition sentences, in which we look for faceted terms that represent the NE's class properties. The reason for relying on the contents of the Wikipedia articles for deriving the NEs facets instead of exploring the Wikipedia categories for those NEs, summarize to the following. First, not all Wikipedia articles about NEs have been associated with a category label. Moreover, most of the articles about NEs that are topically annotated have been assigned to numerous categories, whose class denotations and interrelations are not clearly distinguishable. Lastly, facets unlike topics represent the class properties of a concept (i.e. NE) rather than its semantic orientation. Therefore, in our work we decided to select the facets that describe the NEs class properties from the terms collectively selected by the Wikipedia editors for clarifying and/or defining NE.

Therefore, we address the challenge of identifying useful facets for representing the NE properties as a term extraction problem for which we need a sound model that accurately detects valuable faceted terms within the NEs' surrounding context. The greatest difficulty associated with implementing such a model is how to acutely detect which of the NEs' contextual features are good facets for describing the entity's properties. By good facets, we denote the terms that help us discriminate between the different named entity references while at the same time they are not overly specific, they are not redundant across different entity types and they are useful in a web search setting. To build a model that complies with the above criteria, we have designed a facet extraction algorithm, the basic steps of which are illustrated in Fig.1. In brief, our algorithm takes as input a set of named entities E for which we would like to define useful facets, and a set of clarification and definition sentences S that have been extracted from the Wikipedia pages that correspond to each of the above named entities. Based on the above data and following the below-listed steps, our algorithm

Input: Wikipedia clarification and definition sentences S, set of Named Entities E
Output: useful facet terms (F)
For each s in S do
  Extract all content terms T
      For each t in T **do**
        Compute R(t)
        Associate every t with corresponding entity e_i
      end
end
For each entity e_i in E collect all content terms associated with e_i, $T_i(e_i)$
  /*Compute terminological similarity*/
    For each e_i, e_j with terms Ti, Tj **do**
      $S_{term}(e_i, e_j)$
      If $S_{term}(e_i, e_j) > 0.5$
        Group e_i, e_j together
    end
end
For each e_i in class c
  Take all content terms t associated with e_i
  For each t in c **do**
    Compute R(t,c)
    Compute Usefulness (t)
  end
end
return top k-terms in Facet (F), ranked by Usefulness (t)

Fig. 1. Identifying useful NE faceted terms in the contextual elements of Wikipedia articles

delivers for every class of named entities a sorted list of facets F that are useful for describing the named entity properties within their reference classes.

The first step of our approach extracts all content terms from the clarification and definition sentences (S) that have been collected for each of the named entities. For content terms extraction, we apply tokenization and Part-of-Speech tagging to all the sentences contained in S and we retain only the sentence nouns and proper nouns, based on the observation of [19] that terms of the above grammatical categories communicate most of the thematic properties of the text in which they appear. Then, for every extracted content term t, we estimate the fraction of named entities that contain t in their definition and clarification sentences. For our estimations, we firstly associate every term extracted from a sentence s to the respective entity e that is clarified or defined in the contents of s. Such *content term-named entity* associations can be easily derived from the associations between the named entities and the Wikipedia sentences about the entities that contain t. Then, we compute for every term t a value R(t) that indicates the "representation ratio" of t with respect to some entity, given by:

$$R(t) = | E(t) | / | E | \qquad (1)$$

Where |E(t)| is the count of all named entities that contain t in their clarification and definition sentences and |E| is the count of all named entities identified. The value of R indicates how representative is a term for describing named entity properties, assuming that as the value of R increases so does the term's probability of being a good facet for named entities. At the end of step 1, we associate every named entity with a set of content terms extracted from the Wikipedia sentences that describe the entity properties. Each of the extracted terms is also associated with an overall representation value, i.e. R score.

In the second step we measure the amount of overlapping content terms between named entity pairs, so as to be able to determine for every pair of named entities e_i, e_j whether they refer to the same class of subjects or not. In this respect, we rely on the content terms associated with every named entity (in step 1) and we compute the similarity (S_{term}) between named entity pairs as follows:

$$S_{term}(e_i, e_j) = \frac{2 \bullet |\text{common terms}|}{|\text{terms about } e_i| + |\text{terms about } e_j|} \quad (2)$$

The similarity between the named entities (e_i, e_j) content terms, takes values between 0 and 1; with zero indicating that the two entities have no content terms in common in their clarification and definition sentences and one indicating that all the terms in the named entity sentences are common.

Based on the above formula, we can derive for any pair of named entities the degree to which their contextual elements overlap. We then determine whether any two named entities refer to the same class of subjects based on the following criterion:

$$e_i, e_j = \begin{cases} \text{same class} & \text{if } S_{term}(e_i, e_j) > 0.5 \\ \text{different class} & \text{otherwise} \end{cases} \quad (3)$$

This way, we group named entities into subject classes according to their content terms' similarity values. Note that under our criterion, a named entity e_k might be grouped under several classes depending on the number of named entities with which e_k shares a significant amount of content terms. Having grouped NEs according to their contextual overlapping features, the next step is to determine a set of terms for representing the subject denotations of the named entities that are grouped together. More specifically, we need to identify among the content terms that have been determined for each of the NEs that refer to the same class, the ones that make the most useful and descriptive facets for representing the NEs' properties within that class.

In the last step, our algorithm determines useful faceted terms for each of the named entity classes as follows. Given a set of named entities that refer to the same subject class (i.e. they are grouped together) it collects all their content terms and starts by estimating a new R value for every term within a class, as follows:

$$R(t,c) = |E(t,c)| / |E(c)| \quad (4)$$

Where |E(t, c)| is the count of all named entities that refer to class c and which contain t in their clarification and definition sentences and |E(c)| is the count of all the named entities identified for class c. This new R(t,c) value indicates the "representation ratio" of t with respect to some class c so that terms with increased R(t,c) are better candidates for representing the class properties and consequently the NEs that refer to that class. Having computed the degree to which a content term t represents the properties of the entities that refer to a given class (i.e. R(t,c) value) and considering that we have already estimated (in step 1) the degree to which t makes a good term for representing NE properties, we easily derive the usefulness of t as a facet, as:

$$\text{Usefulness (t)} = R(t) \bullet R(t, c) \tag{5}$$

Based on the above formula, we estimate the usefulness of a term t in serving as a facet for some named entity that refers to a class c as the product of the term's representation ratio for named entity properties and the term's representation ratio for subject class properties. At the end of this process, we retain the top-k terms in each of the subject classes as the faceted terms that are useful in denoting the properties of the named entities that refer to that class.

4   Faceted Search for Named Entity Queries

So far we have presented our method towards identifying both the number of classes to which a named entity query refers and a set of useful facets for representing the query properties within every identified class. One last issue that our study addresses is how to be able to identify the query subject denotations within the contents of the retrieved pages. In this respect, we suggest an approach that examines the distribution of the query facets in the contents of the query retrieved pages and groups search results according to their query denotation types. In particular, given a NE query and a set of facets identified for each of the referring query classes our method employs a simple string matching approach and looks for the query faceted terms in the contents of the search results. In case a page contains some of the query facets, our module investigates whether these pertain to one or more subject classes. If all detected query facets represent a single class, then the faceted term of the highest usefulness value for that class is selected as the facet that represents the NE query denotation in the pages contents. The selected facet serves as a tag that is displayed next to the page in the search engine results. Conversely, if the query facets that a page contains represent multiple classes, then our module examines the position of the identified facets with respect to the query keywords in the page's content and selects the facet that is closest to the query terms as the one that represents the subject denotation of the NE query in the page. Again, the selected facet is used as a tag that indicates the query class representation in the page's content. Finally, in case a page does not contain any of the query facets, it receives no tag and as such it cannot inform the user about the NE query denotations. However, in the latter case, we suggest that query facets are displayed together with search results and enable the user click on the facet term that best suits her query intention. The selected facet is then appended to the query keywords and the refined query is re-submitted to the engine. Following the annotation of the query retrieved pages with an appropriate faceted term from the ones that have been identified for a NE query, we suggest grouping search results by subject denotations (i.e. faceted terms) and display them to the users accompanied by their identified tags. Based on this enriched list of search results, the users can make informed clicking decisions and satisfy their information needs faster.

5   Experimental Evaluation

In this section, we present the experimental evaluation of our facet extraction algorithm and we discuss obtained results. Due to the absence of a standard benchmark for evaluating the usefulness of the automatically selected NE facets, we carried out a human study, in which we measured the accuracy of the facets identified by our algorithm. For our experiment, we relied on 7,000 randomly selected named entities and their corresponding Wikipedia articles that served as our experimental data. Given that some of the NEs are discussed in more than one Wikipedia articles, the total set of the NE pages that we examined in our study is 19,350.

Following the method presented in Section 3.1, we processed the above data in order to determine for every NE the number of referenced subject classes. From all the NEs in our dataset, 410 had a single reference class and the remaining 6,590 had multiple reference classes (between 2 and 7). Then, we relied on the Wikipedia clarification and definition sentences that we extracted from the articles discussing each of our experimental NEs and we supplied them as input to our facet selection algorithm. The latter, following the steps discussed in Section 3.2, grouped the NEs into subject classes and for every class it identified a set of useful facets for denoting the NE properties within that class. In total our algorithm computed 12 subject classes for grouping our experimental NEs and for every class it selected k ($k=10$) faceted terms (i.e. a total set of 120 facets). The most useful facets in each of the identified classes are: *Person, Institute, Holiday, Country, Corporation, Novel, Disease, Newspaper, Science, Film, Band* and *War*.

To asses the performance of our algorithm in selecting useful NE facets, we carried out a human study in which we evaluated the accuracy of the facets that our algorithm selected for denoting the NE reference classes. To conduct our study, we recruited 45 volunteers from our school to whom we presented the list of NEs and their corresponding 19,350 Wikipedia pages and asked them for every NE to read the respective pages and indicate a set of terms that was in their opinion useful for representing the subject denotation of the NE within every page. We asked our participants to select up to 10 terms for every page referring to a NE and we indicated that terms may or may not appear in the contents of that page. Each of the 19,350 Wikipedia pages was examined by five participants and we considered a manually defined faceted term to be valid if at least three of the participants selected the same term to represent the NE reference in the page. In total, our participants identified 269 distinct facets of which 204 were selected by at least three different users. We then relied on these 204 jointly selected distinct facets and we compared them to the 120 facets that our algorithm selected for the same set of named entities and reference pages.

For our comparisons, we relied on the OSim measure [20] and computed for every NE the degree of overlapping facets between those selected by our participants and those delivered by our algorithm, Formally, the overlap between two lists of facets (each of size k) for a given NE is determined as:

$$\text{OSim}(F_{manual}, F_{system}) = \left| T_{manual} \cap T_{system} \right| / k \qquad (6)$$

Where T_{namual} is the set of NE faceted terms that our subjects indicated, T_{system} is the set of NE faceted terms that our algorithm selected and k is the number of facets considered, which in our case $k=10$ since both our participants and our algorithm delivered up to 10 facets for every NE. Table 1 lists obtained results. Due to space constraints, we report the number of our experimental named entities that exhibit similar degrees of overlapping facets across the different OSim levels. As the Table shows, our algorithm managed to identify for 95% of the NEs (i.e. for 6,650 out of the 7,000) at least one facet that was identical to a human selected one. A close look at the obtained results reveals that for NEs with highly overlapping facets, our participants picked faceted terms from the contents of the named entity pages. This demonstrates our algorithm's ability in identifying within the NE contextual elements the terms that are highly representative of the named entity denotations.

Although OSim is a useful measure for deriving the level of agreement between the human and the system selected facets, nevertheless it is marginally informative about how people perceive the usefulness of the automatically selected NE facets. To be able to judge that, we asked our participants to examine the facets selected by our algorithm and indicate which of these (if any) were in their opinion useful for representing the corresponding NE references in the contents of their pages. Human judgments on the usefulness of the automatically selected facets were binary (i.e. useful, non-useful) and as in the case of OSim, every facet was examined by five participants and we considered a facet to be useful if at least three of the users marked the facet as such. We list obtained results in Table 2.

Table 1. NEs distribution at overlapping similarity levels between the user-defined and the system-selected facets

#of Named Entities	% of overlapping facets
350	0
785	0.1-0.2
806	0.2-0.3
1,965	0.3-0.4
1,622	0.4-0.5
778	0.5-0.6
424	0.6-0.7
170	0.7-0.8
80	0.8-0.9
20	0.9-1

Table 2. Statistics on the human judgments about the usefulness of the automatically selected facets

# of algorithm selected facets	120
# of useful facets	86
# of non-useful facets	34
algorithm's success rate	71.7%

As Table 2 demonstrates most of the facets that our algorithm selects are deemed as useful by our participants, yielding an overall 71.7% algorithm success rate in automatically selecting useful named entities facets.

Overall, the results of our human study indicate that users consider the facets that our algorithm selected for representing the NE reference classes as useful. Given our algorithm's potential, we believe that it can be fruitfully employed in a web search setting in order to represent the named entity properties within both the user queries and the query retrieved pages. By using facets to represent both the query and the pages' subjects, we believe that users will be able to locate pages of interest faster and conveniently. In this respect, and considering that NE facets can be computed offline, we deem that our method can operate on-the-fly and be readily explored in web search applications.

6 Concluding Remarks

We presented a method that automatically identifies terms in the contextual elements of NE queries that are representative of the query subject denotations. We have also introduced a metric that estimates the usefulness of every identified facet for a given query reference class. Our experimental evaluation indicates that the facets selected by our algorithm are useful for denoting NE properties and as such we believe that they can be employed towards improving web searches about NEs. Currently, we are testing our algorithm's efficiency on a different corpus in order to validate how much the dataset used for the extraction of NE facets influences the quality of the obtained results.

References

1. Li, X., Liu, B., Yu, P.: Mining Community Structure of Named Entities from Web Pages and Blogs. In: AAAI Spring Symposia on Computational Approaches to Analyzing Weblogs (2006)
2. Pasca, M.: Weakly-Supervised Discovery of Named Entities Using Web Search Queries. In: Proceedings of the 16th ACM Conference on Information and Knowledge Management (2007)
3. Cucerzan, S., Yarowsky, D.: Language Independent NER Using a Unified Model if Internal and Contextual Features. In: Proceedings of CoNLL Conference, pp. 171–174 (2002)
4. Klein, D., Smarr, J., Nguyen, H., Manning, C.D.: Named Entity Recognition with Character Level Models. In: Proceedings of the CoNLL Conference (2003)
5. Fleischman, M., Hovy, E.: Fine Grained Classification of Named Entities. In: Proceedings of the COLING Conference, pp. 267–273 (2002)
6. Florian, R., Ittycheriah, A., Jing, H., Zhang, T.: Named Entity Recognition through Classifier Combination. In: Proceedings of the CoNLL Conference, pp. 168–171 (2003)
7. Bunescu, R., Pasca, M.: Using Encyclopedic Knowledge for Named Entity Disambiguation. In: Proceedings of the EACL Conference, pp. 9–16 (2006)
8. Dakka, W., Cucerzan, S.: Augmenting Wikipedia with Named Entity Tags. In: Proceedings of the 3rd Intl. Joint Conference on Natural Language Processing (2008)
9. Watanabe, Y., Asahara, M., Matsumoto, Y.: A Graph-Based Approach to Named Entity Categorization in Wikipedia Using Conditional Random Fields. In: Proceedings of the EMNLP-CoNLL Conference, pp. 649–657 (2007)
10. Dakka, W., Dayal, R., Ipeirotis, P.: Automatic Discovery of Useful Facet Terms. In: Proceedings of the ACM SIGIR Workshop on Faceted Search (2006)
11. Dakka, W., Ipeirotis, P.: Automatic Extraction of Useful Facet Hierarchies form Text Databases. In: Proceedings of the ICDE Conference (2008)
12. Fellbaum, C.: WordNet: An Electronic Lexical Database. MIT Press, Cambridge (1998)
13. Cucerzan, S.: Large-Scale Named Entity Disambiguation Based on Wikipedia Data. In: Proceedings of the EMNLP Conference (2007)
14. Nguyen, B.V., Kan, M.-Y.: Functional Faceted Web Query Analysis. In: The WWW 2007 (2007)
15. Koren, J., Zhang, Y., Liu, X.: Personalized Interactive Faceted Search. In: The WWW 2008 (2008)
16. Tunkelang, D.: Dynamic Category Sets: An Approach for Faceted Search. In: Proceedings of the ACM SIGIR Workshop on Faceted Search (2006)

17. Di, N., Yao, C., Duan, M., Zhu, J.J.-H., Li, X.: Representing a Web page as Sets of Named Entities of Multiple Types – A Model and Some Preliminary Applications. In: Proceedings of the World Wide Web Conference (poster session), pp. 1099–1110 (2008)
18. Dash, D., Rao, J., Megoddo, N., Ailamaki, A., Lohman, G.: Dynamic Faceted Search for Discovery-Driven Analysis. In: Proceedings of the 17th Intl. ACM CIKM Conference (2008)
19. Gliozzo, A., Strapparava, C., Dagan, I.: Unsupervised and Supervised Exploitation of Semantic Domains in Lexical Disambiguation. Computer Speech and Language 18(3), 275–299 (2004)
20. Haveliwala, T.: Topic Sensitive PageRank. In: Proc. of the 11th WWW Conference (2002)

A Petri-Net-Based Approach to QoS Estimation of Web Service Choreographies*

Yunni Xia[1], Jun Chen[2], Mingqiang Zhou[1], and Yu Huang[1]

[1] School of computer science of Chongqing Univ., China
xiayunni@yahoo.com.cn
[2] School of software of Chongqing Univ. of Posts and Telecommunications, China

Abstract. Current Web service choreography proposals, such as WSCI and BPEL, provide notations for describing the message flows in Web service collaborations. The kernel of WSCI consists of simple communication primitives that may be combined using control-flow constructs expressing sequence, branching, parallelism, synchronization, etc. Many efforts have been made on functional formalization and property verification of WSCI-based service choreography. However, QoS facet of service is yet to be given the importance it deserves. In this paper, we introduce a novel analytical approach to predict the QoS of web service choreographed based on WSCI using general stochastic Petri net (GSPN) as the intermediate representation. From the GSPN model and its corresponding continuous-time Markov chain, analytical estimation of three QoS metrics are obtained. In the case study, we also use computer simulation and confidence interval analysis to validate theoretical evaluations.

1 Introduction

The idea of service WSCI[1] is a composition language that describes the messages exchanging between Web services that participate in a collaborative exchange. A key aspect of WSCI is that it describes only the observable behavior between Web services using temporal and logical dependencies among the exchanged messages and featuring sequencing rules, correlation, exception handling and transactions.

Many efforts have been made on functional formalization and property verification of WSCI-based service choreography. Researches based on process algebra [2,3,4], Petri net[5,6,7] and automata[8,9] are introduced recently to formally capture behavioral patterns of WSCI and build upon a large body of theoretical results as well as techniques for verifying properties such as liveness, soundness, substitutability, boundness and fairness.

However, less attention is paid on quality-of-service facet of service composition. Due to the complexity of the problem, most contributions employ testing, bench-marking or simulation (simulation in our research is merely used as a

* Supported by national 863 plans projects of China under grant number NO.2006AA10Z233.

L. Chen et al. (Eds.): APWeb and WAIM 2009, LNCS 5731, pp. 113–124, 2009.

supplementary tool to validate theoretical models) techniques as their means for QoS evaluation. For instance, [10] develops a bench-marking tool named JWSPerf to obtain run-time performance records. [11,12] introduce a testing bed for performance evaluation through extracting performance-related information from SOAP message exchanges. [13] introduces a method to augment web service composition descriptions to support QoS documentation and analyzes QoS based on run-time logs. [14] uses computer simulation to analyze workload of WS-BPEL based business processes. On the other hand, analytical approaches for QoS evaluation are even more limited and preliminary. For instance, [15] proposes a method to interpret BPEL descriptions into stochastic Petri net and derives reliability estimates. However, its interpretation method is coarse-grained and only captures structural activities. The interpretation from BEPL into UML introduced in [16] is similarly coarse-grained. [17] introduces a structural reduction approach for performance evaluation of web service processes but requires activities to be with deterministic execution duration.

In this paper, we introduce a stochastic approach to evaluate QoS of WSCI-based service choreography employing the general stochastic Petri net (GSPN) as the intermediate model. We first introduce translation rules to map activities, routing patterns(sequence, all, choice, switch, while) and other constructs(handler, timer) into GSPN fragments. Based on probabilistic transition matrix derived from GSPN, we present analytical methods to evaluate three QoS metrics (expected-process-normal-completion-time, process-normal-completion-probability and expected-overhead-of-normal-completion). We also use simulation and confidence interval analysis to validate theoretical results by showing theoretical evaluations of metrics are perfectly covered by corresponding 95% confidence intervals derived from simulative results.

2 Preliminaries

2.1 WSCI

WSCI addresses Web services composition from two primary levels. At the first level, WSCI builds up on the WSDL (Web Service Description Language) *portType* capabilities to describe the flow of messages exchanged by a Web service. The interface construct introduced by WSCI permits the description of the externally observable behavior of a Web service, facilitating the expression of sequential and logical dependencies of exchanges at different operations in WSDL *portType* element. At the second level, WSCI defines the model construct which allows choreography of two or more WSCI interface definitions (of the respective Web services) into a collaborative process involving the participants represented by the Web services. WSCI calls this global model which provides a global, message-oriented view of the overall process.

2.2 Stochastic Petri Net

Petri Nets is a tool used for modeling and analysis of complex system with behavioral patterns such as concurrency, synchronization and conflict. Original

Petri net does not care the concept of time and is extended into various types of timed/stochastic Petri nets. We base our research on general stochastic petri net (GSPN)[18]:

Definition. A GSPN is a 4-tuple (P, T, F, λ):

1. $P = \{p_1, p_2, ...p_n\}$ is a finite set of places
2. T is a finite set of transitions partitioned into two subsets: TI (immediate) and TD(timed) transitions. Immediate transitions are depicted by solid bars while timed transition by hollow bars
3. $\lambda : TD \rightarrow real$ is a function identifying execution rate of each timed transition
4. $F \subseteq (P * T) \cup (T * P)$ is a set of directed arcs

3   Transforming WSCI into GSPN

In this section, we illustrate the translation process from WSCI activities and constructs (following the syntax defined in Figure.1 given by [3] which is isomorphic to WSCI but without the verbosity of the XML tags) into GPSN fragments.

The first translation rule deals with message exchanges. WSCI atomic actions correspond to input and output actions on the corresponding message channels. A message is represented by a token and the arrival of a message in a specific channel represented by the existence of one token in the corresponding GSPN place, whose name is formed by the *portType*, operation and message names (e.g.,*port/op/msg*). The translation is illustrated in Fig.2(a).

```
interfaceDef ::= interface name = { processDef⁺ }
processDef ::= process name = { [contextDef ] activityDef⁺ }
contextDef ::= context { processDef*[exceptionDef ] }
activityDef ::= actionDef
              | sequence [name] { [contextDef ] activityDef⁺ }
              | all [name] { [contextDef ] activityDef* }
              | switch [name] { case⁺[default ] }
              | choice [name] {onMessage*| onTimeout*| onFault*}
              | foreach [name] { list } do { [contextDef ] activityDef⁺ }
              | while [name] { boolExpr } do { [contextDef ] activityDef⁺ }
              | until [name] { boolExpr } do { [contextDef ] activityDef⁺ }
              | empty [name]
              | fault [name] { faultName }
              | call [name] { processName }
              | spawn [name] { processName }
actionDef ::= operation [name] operation ; [call [name] { processName } ; ]
operation ::= oneWay | requestResp | notification | solicitResp
oneWay ::= in msg
requestResp ::= in msg ; out msg
notification ::= out msg
solicitResp ::= out msg ; in msg
msg ::= wsdlPortType/wsdlOperation/msgName
case ::= case { boolExpr } do { [contextDef ] activityDef⁺ }
default ::= default { [contextDef ] activityDef⁺ }
exceptionDef ::= exception {onMessage*| onTimeout*| onFault*}
onMessage ::= onMessage { (oneWay | requestResp) [contextDef ] activityDef* }
onTimeout ::= onTimeout { timeout } do { [contextDef ] activityDef⁺ }
onFault ::= onFault { name } do { [contextDef ] activityDef⁺ }
timeout ::= fromStartOf name duration timeDuration
           | fromEndOf name duration timeDuration
```

Fig. 1. Grammar for the untagged version of WSCI

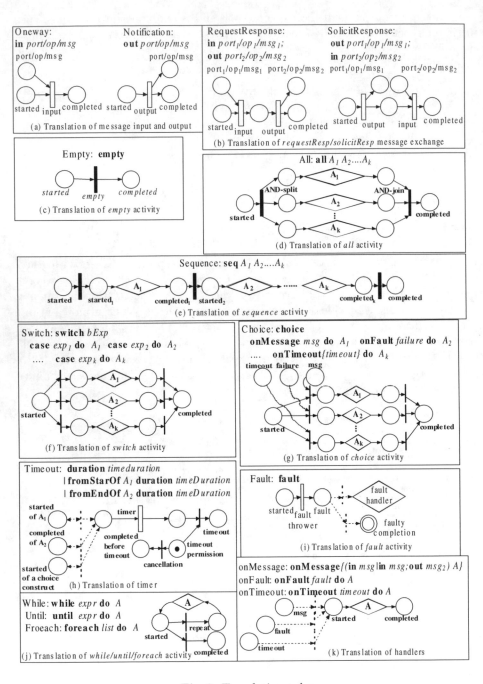

Fig. 2. Translation rules

Since the *requestResp* operation can be viewed as the concatenation of a message input and a message output, while the *solicitResp* operation can be viewed as the concatenation of a message output and a message input, their translation is given by Fig.2(b).

Obviously, the *empty* action can be translated into a single immediate transition, as illustrated in Fig.2(c).

The mapping rule of *all* construct is illustrated in Fig.2(d), where included activities (may be atomic or complex denoted by diamonds) are activated in parallel using the $AND - split$ and $AND - join$ control structures.

The translation of *sequence* construct is straightforward and given by Fig.2(e).

Although both *switch* and *choice* capture nondeterministic activation of branches, there selection mechanisms are intrinsically different. The decision-making of *switch* is internally(externally unobservable) determined by the truth value of boolean expression associated with each branch. Consequently, the translation employs the $XOR - split/XOR - join$ routing structure to express the internally nondeterministic decision-making. On the other hand, branch selection of *choice* construct is externally decided by messages, timeout signals or faults. The translation must therefore explicitly capture influences by those external factors. Translation rules of the two constructs are illustrated in Fig.2(f,g).

The timer of WSCI is translated into GSPN fragment shown in Fig.2(h). According to the grammar of WSCI, if no $fromStartOf/fromEndOf$ condition is imposed, a timer is activated when the *choice* activity (which is associated with that time) is started. The activation can also happen when a specified activity is started or completed using the $fromStartOf/fromEndOf$ condition. A timer can be canceled(through moving away the default token in $timeout-permission$ place in Fig.2(h)) when the *choice* construct or the most inner construct including the context of that timer(if $fromStartOf/fromEndOf$ condition is imposed) is completed before timer expiration. In Fig.2(h), dashed arrows and transitions means they may be absent.

Translation rule of the *fault* activity is illustrated in Fig.2(i). The *fault* activity is an atomic activity that triggers a fault in the current context. If the fault is not handled by any exception event handler defined in the current context, or in any parent context, the entire WSCI process will complete at the faulty status(denoted by the $faulty - completion$ place in Fig.2(i)).

$until/while/foreach$ activities repeatedly execute their including activities and the translation is given by Fig.2(j).

$onMessage/onTimeout/onFault$ handlers are activated when exceptional message, timeout message or fault occur. Their translations are illustrated by Fig.2(k).

4   An Example

In this section, a choreography sample is given in Fig.3 and translated into GSPN using translation rules given in the previous section. The sample is an modified version of the sample used [1] and captures a airline-ticket-booking application. The choreography includes two interfaces, namely *traveler* and $travel - agent$.

```
<interface name = "Traveler">
 <process name = "PlanAndBookTrip" instantiation = "other">
  <sequence>
   <context>
    <exception>
     <onMessage>
      <action name = "UnavailabilityHandling"
            role = "tns:traveler"
            operation = "tns:TravelerToTA/BookingCLosed">
       <faultcode = "tns:exit"/>
      </onMessage>
      <onMessage>
       <action name = "TimeOutHandling" role = "tns:traveler"
            operation = "tns:TravelerToTA/TimeOut"/>
        <faultcode = "tns:exit"/>
      </onMessage>
    </exception>
   </context>
   <sequence>
    <action name = "BookingNotification" role = "tns:traveler"
         operation = "tns:TravelerToTA/BookingNotification"/>
    </action>
    <action name = "AvailabilityConfirm" role = "tns:traveler"
         operation = "tns:TravelerToTA/BookingAvailable"/>
    </action>
   <switch>
```

```
<case>
 <condition>tns:ItineraryA</condition>
  <action name = "BookTicketA" role = "tns:traveler"
       operation = "tns:TravelerToTA/BookTicketA"/>
</case>
<default>
 <action name = "BookTicketB" role = "tns:traveler"
      operation = "tns:TravelerToTA/BookTicketB">
 </action>
</default>
</switch>
<action name = "BookingConfirm" role = "tns:traveler"
     operation = "tns:TravelerToTA/BookingConfirm">
<all>
 <action name="ProvideAccount" role="tns:traveler"
      operation="tns:TravelerTOTA/ProvideAccount"/>
 <while name="ReserveSeats">
  <condition>defs:notLastSeat</condition>
   <action name="ReserveSeat" role ="tns:Traveler"
        operation="tns:TravelerToTA/ReserveSeat"/>
 </while>
</all>
</sequence>
</process>
</interface>
```

(a)The interface of traveler

```
<interface name = "TravelAgent">
 <process name = "TestAvailability" instantiation = "other">
  <switch>
   <case>
    <condition>tns:Unavailable</condition>
    <action name = "BookingClosed" role = "tns:travelAgent"
         operation = "tns:TAtoTraveler/BookingClosed"/>
   </case>
   <default>
    <action name = "BookingAvailable" role = "tns:travelAgent"
         operation = "tns:TAtoTraveler/BookingAvailable"/>
   </default>
  </switch>
 </process>
 <process name="AccountRetrieval" instantiation="message">
  <action name="RetrieveAccount" role="tns:travelAgent"
       operation="tns:TAtoTraveler/ProvideAccount"/>
 </process>
 <process name="ReserveSeat" instantiation="message">
  <action name="ReserveSeatConfirm" role="tns:travelAgent"
       operation="tns:TAtoTraveler/ReserveSeat"/>
 </process>
 <process name = "PlanAndBookTrip" instantiation = "message">
  <sequence>
   <sequence>
    <context>
```

```
<exception>
 <onTimeout property = "tns:expiryTime"
      type = "duration"
      reference="tns:ReceiveTripOrder@start">
  <action name="NotifyTimeout" role="tns:travelAgent"
       operation="tns:TAtoTraveler/Timeout"/>
  </onTimeout>
 </exception>
</context>
<action name = "ReceiveTripOrder" role = "tns:travelAgent"
     operation = "tns:TAtoTraveler/BookTickets"/>
 <spawn process = "TestAvailability"/>
 <join process="TestAvailability"/>
</action>
</sequence>
<choice>
 <onMessage>
  <action name = "ReceiveA" role = "tns:travelAgent"
       operation = "tns:TAtoTraveler/BookTicketA"/>
  <action name="ConfirmBooking" role="tns:travelAgent"
       operation="tns:TAtotraveler/ConfirmBooking"/>
 </onMessage>
 <onMessage>
  <action name = "ReceiveB" role = "tns:travelAgent"
       operation = "tns:TAtoTraveler/BookTicketB"/>
  <action name="Confirm Booking" role="tns:travelAgent"
       operation="tns:TAtotraveler/ConfirmBooking"/>
 </onMessage>
</choice>
</sequence>
</process>
</interface>
```

(b)The interface of travel-agent

Fig. 3. The sample of service choreography

All message exchanges are of single direction with type *oneWay* or *notification*. For brevity, *portType* definitions, correlation definitions and global model is omitted.

The traveler interface captures booking behaviors of clients, including a main process *PlanAndBookTrip*. The process first sends the travel-agent a notification signal to start the booking application through *BookingNotification* and

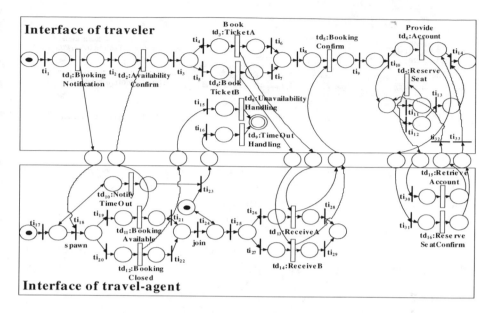

Fig. 4. The GSPN derived from the sample choreography

waits for reply through *AvailabilityConfirm*. If the reply message indicates unavailability of service, *UnavailabilityHandling* in the exception definition is activated and the handler throws a fault to terminate the application. On the other hand, on receipt of positive reply, the traveler starts a internal nondeterministic decision between two optional ticket offerings. After all mentioned above are accomplished, the traveler finally starts two parallel branches to provide account information and reserve multiple seats.

The travel-agent interface captures booking service of ticket provider, including a main process *PlanAndBookTrip* (according to [1], it is recommended to assign the same name to all top processes which belong to the same application). The process first reply to *BookingNotification* of traveler through *ReceiveTripOrder* and spawns the *TestAvailability* process defined in its context. The *TestAvailability* process may reply with availability or unavailability status. When *ReceiveTripOrder* starts, a timer is activated and the timer throws a timeout exception if it takes *ReceiveTripOrder* too long to reply. After *ReceiveTripOrder* ends, a choice activity is started where only one of *ReceiveA* and *ReceiveA* is executed according to message received from the traveler. The interface also includes two message-driven processes *AccountRetrieval* and *ReserveSeat* which respectively respond to *ProvideAccount* and *ReserveSeat* of the traveler interface.

Based on translation rules given in the previous section, the sample choreography can be translated into the GSPN given in Fig.4, where the upper and lower part respectively illustrate the traveler and travel-agent interface and message channels are mapped into places between the two interfaces. The place with double circles denotes the status of faulty-completion. Initially, there is each one

Table 1. OPUT of timed transitions

TD	td_1	td_2	td_3	td_4	td_5	td_6	td_7	td_8	td_9	td_{10}	td_{11}	td_{12}	td_{13}	td_{14}	td_{15}	td_{16}
λ	0.74	0.63	1	1.2	0.88	0.41	0.39	0.94	0.65	0.57	0.29	0.46	0.71	0.55	1.08	0.38
OPUT	1.23	0.89	0.65	1.04	1.7	2.53	0.48	0.93	2.75	1.46	0.53	0.94	0.58	1.33	0.72	0.42

token in the starting place of the traveler process, the starting place of the agent process and activation place of the timer construct (according to the translation rule given in Fig.2(h)).

For QoS analysis, we have to define two more functions. $pe : TI \rightarrow Real$ identifies the probability that each every immediate transition is executed when its preceding places contain tokens. Since only $ti_{4/5/11/12/19/20}$ are nondeterministically activated, we have that $pe(ti_{4/5/11/12/19/20}) = 0.2/0.8/0.5/0.5/0.95/0.05$ and pe for all other immediate transitions equal 1. Another function is $OPUT : TD \rightarrow Real$ to identify over-head-per-unit-time of every timed transition. Execution rate and $OPUT$ of every timed transition of the sample is given in Table.1.

5   Stochastic Modeling and QoS Evaluation

For quantitative evaluation of service choreography, a state-based analysis of GSPN is inevitable. Let $X(t)$ denote the set of operational timed transitions at time t (execution begins at time 0), then its state-space S can be obtained in a traversal way. The state-space of the choreography sample is given in Table.1 (For brevity, only a part of the state space is given). According to Fig.4, $s_{6,8,9}$ can only leads to faulty completion since these states indicate activation of handlers and all handlers in the sample throws faults which can not be caught.

Table 2. State space

state	operational timed transition	state	operational timed transitions
s_1(Initial-state)	$\{td_1\}$	s_2	$Fauty\ completion$
s_3	$Normal\ completion$	s_4	$\{td_{10}, td_{11}\}$
s_5	$\{td_{10}, td_{12}\}$	s_6	$\{td_9, td_{11}\}$
s_7	$\{td_2, td_{10}\}$	s_8	$\{td_9, td_{12}\}$
s_9	$\{td_8, td_{10}\}$	s_{10}	$\{td_3, td_{10}\}$
$\cdots$			

Since execution duration of every timed transition in GSPN is exponential , we have that $X(t)$ is a homogenous continuous Markov chain and its infinitesimal generator matrix Q is given by Eq.1.

$$q_{i,j} = \begin{cases} \lambda(td_l) \times \prod_{ti_m \in TISET} pe(ti_m) & \text{if } s_i \xrightarrow{td_l, TISET} s_j \\ -\sum_{1 \leq r \leq |S|, r \neq i} q_{i,r} & \text{if } i = j \\ 0 & \text{else} \end{cases} \quad (1)$$

where $\lambda(td_l)$ denotes execution rate of transition td_l, $|S|$ denotes the number of states in the state space and $q_{i,j}$ denotes the transition rate from state s_i to s_j. Relation $s_i \xrightarrow{td_l, TISET} s_j$ implies that s_j is the resulting state of s_j if timed transition td_l and the set of immediate transitions $TISET$ fire. Those resulting states are viewed as different **types** in the Markovian chain according to the **phase-type** property. The proof of Eq.1 is omitted in this paper.

Based on the transition matrix, three Qos metrics of service choreography can be obtained. The first metric is **expected-process-normal-completion-time**($EPNCT$ for simple). From the perspective of state transition, $EPNCT$ denotes the expected duration for initial state to reach normal termination. Note that, execution of handlers are treated not as faulty completion but as normal process evolution since they are explicitly specified by WSCI description and can not cause unanticipated system behaviors.

In conjunction with $EPNCT$, we introduce two more metrics related to the notion of normal completion, i.e., **process-normal-completion-probability** ($PNCP$ for simple) and **expected-overhead-of-normal-completion**($EONC$ for simple). $PNCP$ denotes the probability that process finally reaches the normal termination state. While $EONC$ belongs to the cost dimension of QoS. Process overhead is determined by execution durations of involved timed transitions and overhead-per-unit-time of those transitions.

To evaluate $EPNCT$, we first have to evaluate expected duration for each state to reach the absorbing state of normal termination, $EDT(i)$. We have

$$
EDT(i) = \begin{cases} 0 & \text{if } s_i \text{ is the normal completion state} \\ \infty & \text{if } s_i \text{ is an faulty completion state} \\ \infty & \text{if all immediately succeeding states of } s_i \text{ have } EDT \text{ of } \infty \\ \frac{1}{-q_{i,i}} + \sum_{1 \leq k \leq |S|, k \neq i, EDT(k) < \infty} \frac{q_{i,k} \times EDT(k)}{TEMP_i} & \text{else} \end{cases}
$$
(2)

where $\frac{1}{-q_{i,i}}$ is the expected elapsing duration of state s_i and $TEMP_i$ is an intermediate variable given by

$$
TEMP_i = \sum_{1 \leq k \leq |S|, k \neq i, EDT(k) < \infty} q_{i,k}
$$
(3)

Eq.2 implies that the EDT of a certain state is simply its expected elapsing duration plus weighted $EDTs$ of its immediately succeeding states(excluding faulty completion states and those which lead to faulty completion states).

Based on observations above, we have that $EPNCT$ is obtained as EDT of the initial state

$$
EPNCT = EDT(1)
$$
(4)

To evaluate $PNCP$, we first have to calculate the probability for each state to reach the normal completion state, $PRN(i)$.

$$PRN(i) = \begin{cases} 1 & \text{if } s_i \text{ is the normal completion state} \\ 0 & \text{if } s_i \text{ is an faulty completion state} \\ 0 & \text{if all immediately succeeding states of } s_i \text{ have } PRN \text{ of } 0 \\ \sum_{1 \leq k \leq |S|, k \neq i, PRN(k) > 0} \frac{q_{i,k}}{TEMP_i} \times PRN(k) & \text{else} \end{cases}$$

$$(5)$$

Similar to Eq.2, Eq.5 is given in an iterative manner. PRN for a certain state is the sum of $PRNs$ of its immediately succeeding states (excluding faulty completion states and those which lead to faulty completion states) multiplied by its transition probabilities to those states.

$PNCP$ is then obtained as

$$PNCP = PRN(1) \tag{6}$$

To evaluate $EONC$, we first have to evaluate an intermediate variable, $EO(i)$, which is the expected overhead for state s_i to reach normal completion.

$$EO(i) = \begin{cases} 0 & \text{if } s_i \text{ is the normal completion state} \\ \infty & \text{if } s_i \text{ is an faulty completion state} \\ \infty & \text{if all immediately succeeding states of } s_i \text{ have } EO \text{ of } \infty \\ \frac{\sum_{td_o \in ACT_i} OPTU(td_o)}{-q_{i,i}} + \sum_{1 \leq k \leq |S|, k \neq i, EO(k) < \infty} \frac{q_{i,k} \times EO(k)}{TEMP_i} & \text{else} \end{cases}$$

$$(7)$$

Therefore, $EONC$ is given by

$$EONC = EO(1) \tag{8}$$

Based on methods introduced above, $EPNCT/PNCP/EONC$ of the sample is calculated as $19.14/0.3203/21.05$.

6   Simulation and Confidence-Interval-Analysis

In this section, we employ Monte-carlo simulation and confidence interval analysis to validate analytical methods introduced in the previous section.

Monte-carlo simulation is a flexible performance prediction tool used widely in science and engineering [19]. In Monte-carlo simulation, stochastic behaviors and events of target system are generated using random number generators. Outputs of simulation procedure are treated as random observations (samples) of the system under study. As for research in this paper, the Monte-carlo simulation procedure conducts 10000 simulation runs of the WSCI sample. In each run, exponential random generators are used to generate execution durations of executed atomic activities and 0/1 random generators are used to decide the choice-making of branches when system encounters nondeterminism. When each simulation run terminates, the final state(normal or faulty completion) is recorded. Also, process-completion-time and process-overhead is calculated and recorded based on execution durations of involved timed transitions.

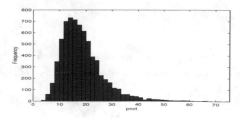
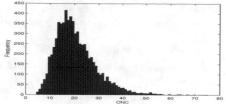

Fig. 5. Hist. chart of simulative $PNCT$ **Fig. 6.** Hist. chart of simulative ONC

Based on histogram charts of simulation experiments illustrated in Fig.5-6, we derive the 95% confidence intervals of $EPNCT/EONC$ as $[18.7650, 19.2073]$ $/[20.8055, 21.2273]$ which perfectly cover theoretical evaluations of 19.14/21.05 given earlier. Using the normal-completion-rate of simulation experiments and the method for confidence interval analysis of **Bernoulli trials** [20], we also obtain the 95% interval of $PNCP$ as $[0.3054, 0.3427]$ which also perfectly covers theoretical evaluation of 0.3203 given earlier. Taken together, results suggest analytical methods in this paper are validated by simulations.

7 Conclusion

In this paper, we introduce a novel analytical approach to predict QoS of service choreography built on WSCI, using GSPN as the intermediate model. From the GSPN model and its corresponding continuous-time Markov chain, analytical evaluation of three QoS metrics are obtained. In the case study, experimental results and confidence interval analysis suggest analytical evaluations are validated by simulation.

References

1. W3C, Web Service Choreography Interface (WSCI) 1.0, World Wide Web Consortium (2002), http://www.w3.org/TR/wsci
2. Liu, F., Shi, Y., Zhang, L., Lin, L., Shi, B.-L.: Analysis of web services composition and substitution via CCS. In: Lee, J., Shim, J., Lee, S.-g., Bussler, C.J., Shim, S. (eds.) DEECS 2006. LNCS, vol. 4055, pp. 236–245. Springer, Heidelberg (2006)
3. Brogi, A., Canal, C., Pimentel, E., Vallecillo, A.: Formalizing Web Service Choreographies. Electr. Notes Theor. Comput. Sci. 105, 73–94 (2004)
4. Xiangpeng, Z., Hongli, Y., Zongyan, Q.: Towards the formal model and verification of web service choreography description language. In: Bravetti, M., Núñez, M., Zavattaro, G. (eds.) WS-FM 2006. LNCS, vol. 4184, pp. 273–287. Springer, Heidelberg (2006)
5. Huang, Y., Xu, C., Wang, H., Xia, Y., Zhu, J., Zhu, C.: Formalizing Web Service Choreography Interface. In: AINA Workshops, vol. (2), pp. 576–581 (2007)
6. Huang, Y., Wang, H.: A petri net semantics for web service choreography. In: SAC 2007, pp. 1689–1690 (2007)

7. Deng, X., Lin, Z., Chen, W., Xiao, R., Fang, L., Li, L.: Modeling Web Service Choreography and Orchestration with Colored Petri Nets. In: SNPD 2007, pp. 838–843 (2007)
8. Diaz, G., Pardo, J.J., Cambronero, M.-E., Valero, V., Cuartero, F.: Verification of Web Services with Timed Automata. Electr. Notes Theor. Comput. Sci. 157(2), 19–34 (2006)
9. Shi, Y., Zhang, L., Liu, F., Lin, L., Shi, B.-L.: Web service collaboration analysis via automata. In: Fan, W., Wu, Z., Yang, J. (eds.) WAIM 2005. LNCS, vol. 3739, pp. 858–863. Springer, Heidelberg (2005)
10. Mashado, A.C.C., Ferraz, C.A.G.: JWSPerf: A Performance Benchmarking Utility with Support to Multiple Web Services Implementations. In: AICT-ICIW 2006, pp. 159–165 (2006)
11. Chen, S., Yan, B., Zic, J., Liu, R., Ng, A.: Evaluation and Modeling of Web Services Performance. In: ICWS 2006, pp. 437–444 (2006)
12. Xie, J., Ye, X., Li, B., Xie, F.: A ConfigurableWeb Service Performance Testing Framework. In: HPCC 2008, pp. 312–319 (2008)
13. McGregor, C., Schiefer, J.: A Framework for Analyzing and Measuring Business Performance with Web Services. In: CEC 2003, pp. 405–412 (2003)
14. Koizumi, S., Koyama, K.: Workload-aware business process simulation with statistical service analysis and timed Petri net. In: ICWS 2007, pp. 70–77 (2007)
15. Zhong, D., Qi, Z.: A Petri net approach for reliability prediction of web services. In: OTM workshop 2006, pp. 116–125 (2006)
16. Zarras, A., Vassiliadis, P., Issarny, V.: Model-Driven Dependability Analysis of Web Services. In: DOA 2004, pp. 1608–1625 (2004)
17. Cardoso, J., Sheth, A., Miller, J., Arnold, J., Kochut, K.: Quality of service for workflows and web service processes. Elsevier Transaction on web semantics 1(3), 281–308 (2004)
18. Balbo, G.: Introduction to generalized stochastic petri nets. In: Bernardo, M., Hillston, J. (eds.) SFM 2007. LNCS, vol. 4486, pp. 83–131. Springer, Heidelberg (2007)
19. Niederreiter, H.: Random number generation, and quasi-Monte-Carlo methods. SIAM, Philadelphia (1992)
20. Leemis, L.M., Trivedi, K.S.: A Comparison of Approximate Interval Estimators for the Bernoulli Parameter. The American Statistician 50(1), 63–68 (1996)

Personalization Service Research of the E-Learning System Based on Clustering Coefficient Partition Algorithm

Hong Liu and Xiaojun Li

College of Computer and Information Engineering
Zhejiang Gongshang University
Hangzhou, 310018, China
{llh,lixj}@mail.zjgsu.edu.cn

Abstract. In recent years, network-based education has been growing rapidly in size and complexity. Therefore, knowledge clustering becomes more and more important in personalized information retrieval for e-learning. This paper introduces a clustering coefficient partition algorithm for providing e-learning personalization service after the learners' knowledge has been classified with clustering. Through automatic analysis of learners' behaviors, their partition with similar knowledge level and interests can be discovered in order to provide learners with contents that best match their educational needs for collaborative learning. Experimental results show that our algorithm is efficient and effective in extracting clusters from large set of contents.

Keywords: e-learning, clustering coefficient, partition, personalization service.

1 Introduction

The rapid development of information technology and the Internet have caused continuous and significant changes to every sector of modern society. Education itself could not remain passive and unconcerned [3,6]: all traditional teaching techniques need re-evaluation, while new ones are introduced. Internet-oriented applications try to satisfy current educational needs by closing the gaps between traditional educational techniques and future trends in technology-blended education. E-learning now becomes a revolutionary way to empower enterprises and their workforce with the necessary skills and knowledge.

Towards this goal, various e-learning systems have been developed in recent years. To address the drawbacks of previous static old-fashioned e-learning system as difficult in doing automatic semantic evaluations, and inefficient in cooperative cognitive processes by providing personalized support such as tracking of learners' input and relevance feedback support for learners. This is particularly important when e-learning takes place within, an open and dynamic learning and information networks. For example, Chen & Chung [1], Chen & Duh [2] and Chen & Liu [7] discussed the applications of item response theory in e-learning system.

L. Chen et al. (Eds.): APWeb and WAIM 2009, LNCS 5731, pp. 125–131, 2009.
© Springer-Verlag Berlin Heidelberg 2009

Cluster analysis is one of the key unsupervised learning techniques of data mining. Clustering techniques have been used for information retrieval, post-searching for categorize information, and knowledge representation for users' understanding. Due to large amounts of education contents continue to grow inexorably in size and complexity, the techniques and approaches of cluster analysis suffer from the challenges such as high-dimensional data clustering, complex data clustering, and description of too many clusters. The purpose of data clustering is to extract the similarity between contents to gain better understanding of them for the purpose of more effective personalized retrieval. Therefore, we apply divisive algorithms for our cluster analysis [8, 9, 10].

The contributions of this paper are below:

1. In this paper, we introduce clustering coefficient partition algorithm into the personalization part of our e-learning system.
2. We propose a system architecture of the E-learning system in which a clustering coefficient partition algorithm is introduced.
3. We further evaluate the performance of our algorithm with experiments.

The rest of paper is organized as follows: In Section 2, we introduce a system architecture. Then, a clustering coefficient partition algorithm is proposed in Section 3. Finally, several experiments are performed to validate the efficiency and effectiveness of this method.

2 System Architecture

As shown in Fig. 1, the architecture of our e-learning system includes three main parts: *learner subsystem*, *instructor subsystem*, and *data management subsystem* [4,5]. These three parts form standalone entities, while at the same time collaborate with one another. The learner subsystem handles both individual learners and learner groups. Individual learner means those participating in the learning process as individuals, while group learners means those participating in the educational process as groups, e.g., collaborative learning. The instructor subsystem handles the instructors, including those from general or special education as well as experts in e-learning. The data management subsystem consists of the personalization engine, system server, and the database.

As shown in Fig. 1, the processes of the learner subsystem include on-line learning, evaluation and coach. The evaluation process produces assessment information and sends it to the coach process. In addition, the evaluation process and coach process create the learner records for storing in the database. Processes of the instructor subsystem include assessment management, queries, learning preference setting, and learning context management. In general, initialization information for learners is defined by the instructors and is stored in the learners' profiles. Each of them describes characteristics of the learner, learning needs, and preferences. The data management subsystem stores knowledge, information, and other resources used in the learning activities in order to integrate the learner subsystem with the instructor subsystem for educational collaboration. The personalization engine introduces a clustering coefficient partition algorithm for analyzing characteristics of learners, learning needs and preferences, and then classifies learner behaviors, and providing new learner personalization services consequently.

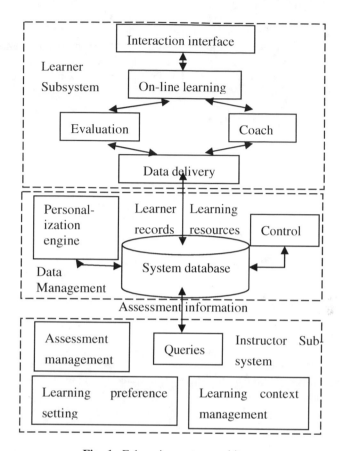

Fig. 1. E-learning system architecture

3 Application of Clustering Coefficient Partition Algorithm in the E-Learning System Personalization Service

The clustering coefficient partition algorithm is a new kind of divisive algorithm that requires the consideration of local quantities only. Therefore it is much faster than GN algorithm. The fundamental ingredient of a divisive algorithm is a quantity that can single out edges connecting nodes belonging to different communities. This is considered as the edge clustering coefficient, defined, in analogy with the usual node clustering coefficient, as the number of triangles to which a given edge belongs, divided by the number of triangles that might potentially include it, given the degrees of the adjacent nodes.

We model a learning behavior information as a directed graph $G=(V,E)$, where V denotes the learner-behavior set, and E represents the route relation set of learners behavior. The clustering coefficient partition algorithm is introduced into the e-learning system personalization service, the concrete steps of which are shown as follows:

(1). Analyze learners' behavior information, and express them as matrix representation of a directed graph, named as a patterned frequency graph that contains a path for each sequence of learning process, as shown in Fig. 2. The rows and columns of the matrix represent learner behavior, while element a_{ij} represents a quantification value of the learner behavior information.

$$\begin{bmatrix} a_{00} & a_{01} & \cdots & a_{0n-2} & a_{0n-1} \\ a_{10} & & & & a_{1n-1} \\ \vdots & & \ddots & \ddots & \vdots \\ a_{m-20} & \ddots & \ddots & \ddots & a_{m-2n-1} \\ a_{m-10} & a_{m-11} & \cdots & a_{m-1n-2} & a_{m-1n-1} \end{bmatrix}$$

Fig. 2. Learners' information matrix

(2). Calculate every edge clustering coefficient $C_{i,j}^{(3)}$, and remove the edge of minimum $C_{i,j}^{(3)}$ in the matrix. We define $C_{i,j}^{(3)}$, in analogy with the usual node clustering coefficient, as the number of triangles to which a given edge belongs, divided by the number of triangles that might potentially include it, given the degrees of the adjacent nodes. More formally, for the edge connecting node i to node j, the edge clustering coefficient is given by:

$$C_{i,j}^{(3)} = \frac{z_{i,j}^{(3)}}{\min[(k_i - 1), (k_j - 1)]} \tag{1}$$

where $z_{i,j}^{(3)}$ is the numbers of triangle built on edge (i,j), and $\min[(k_i-1),(k_j-1)]$ is degree minimum of degree k_i-1 and degree k_j-1.

The symbol $C_{i,j}^{(3)}$ denotes the clustering degree of all the nodes in a graph and shows the link cohesion or *graph tightness*. Adjacent edges based on nodes having weakly adjacent relation are included in little triangle or no triangle with their $C_{i,j}^{(3)}$ small; while there are some triangles based on nodes having tightly adjacent relation with their $C_{i,j}^{(3)}$ relatively large. The idea behind the use of this quantity in a divisive algorithm is that edges connecting nodes in different web graph are included in few or no triangles, and tend to have small values of $C_{i,j}^{(3)}$. On the other hand, many triangles exist within clusters. Hence, the clustering coefficient $C_{i,j}^{(3)}$ is a measure of how inter-communitarian a link is. A problem arises when the number of triangles is zero, because $C_{i,j}^{(3)} = 0$, irrespective of k_i and k_j, (or even $C_{i,j}^{(3)}$ is indeterminate, when min[(k$_i$-1),(k$_j$-1)]=0). To remove this degeneracy, we consider a slightly modified quantity by using, at the numerator, the number of triangles plus one:

$$\tilde{C}_{i,j}^{(3)} = \frac{z_{i,j}^{(3)} + 1}{\min[(k_i - 1), (k_j - 1)]} \tag{2}$$

(3). Judge whether the algorithm get the end, if it is not, repeat the step (2) based on the changed matrix, otherwise, end the algorithm.

In this section, we apply a clustering coefficient partition algorithm in analyzing learners' on-line learning behaviors from their behaviors history. Fig. 3 shows the learners' behaviors, where plus sign denotes one of learners' behaviors having direct connection with another behavior. Fig. 4. shows the corresponding learners' behaviors partition result.

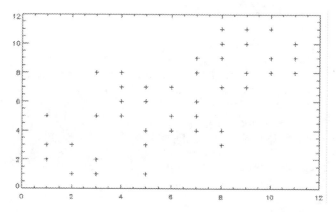

Fig. 3. Learners' behavior before clustering coefficient partition algorithm

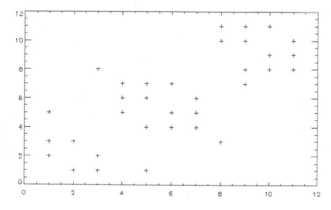

Fig. 4. Learners' behavior partition after clustering coefficient partition algorithm

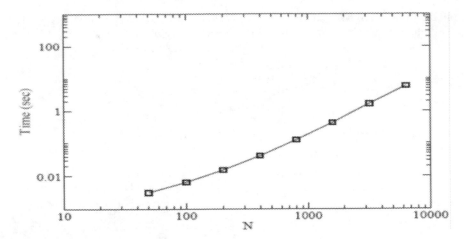

Fig. 5. Efficiency and effectiveness of algorithm

Based on Fig. 4, we demonstrate that such learners' behavior can be divided into three clusters. For a new learner, firstly our system may select a best learner's level that matches, and then based on the result of the partitioning algorithm; our system assigns the best match learning contents, and plans a suitable learning sequence. Then, our system provides the learner with the appropriate e-courses, according to dynamically updated, profiling information.

Fig. 5. shows the efficiency and effectiveness of our algorithm. The computation time of the clustering coefficient partition algorithm is mainly spent in removing edges, because a complete check is necessary to judge whether the algorithm finishes, together with an value update of $C_{i,j}^{(3)}$. The time that the first operation needs is order of N, where N is total edge number of the graph, while the second operation time is constant. So through repeating such operations for all edges, the algorithm time can get to a upper bound of $O(N^2)$. For the graph with a spot of nodes, the relation of compute time with edge number can reduce to linear, conversely, the upper-bound time complexity of the algorithm is $O(N^2)$.

4 Conclusion

This paper presents a clustering coefficient partition algorithm introduced for e-learning personalization. Experiment results demonstrate the efficiency and effectiveness of our proposed method. It analyzes the learner behavior history, after finding out the learner's clustering of knowledge in a specific subject, automatically categorizes to a distinct behavior class that characterizes their behavior as well as future interests and choices within the system. So according to the cluster of learners level to which the learners belong, the e-learning system can provide new learners with appropriate e-courses, educational contents, enabling them to quickly reach the learning goal that is compatible with their knowledge background.

In the future, we will improve the clustering coefficient partition algorithm to get its better efficiency and effectiveness.

Acknowledgment

The work is supported by the Zhejiang provincial department of science and technology as grand science and technology special social development project (No.2008C13082)l The Program of National Natural Science Foundation of China under Grant No.60873022; The Program of Natural Science Foundation of Zhejiang Province under Grant No. Y1080148; The Open Project of Zhejiang Provincial Key Laboratory of Information Network Technology; The Key Project of Special Foundation for Young Scientists in Zhejiang Gongshang University under Grant No. Q09-7.

References

1. Chen, C.-M., Chung, C.-J.: Personalized mobile English vocabulary learning system based on item response theory and learning memory cycle. Computers & Education 51(2), 624–645 (2008)
2. Chen, C.-M., Duh, L.-J.: Personalized web-based tutoring system based on fuzzy item response theory. Expert Systems with Applications 34(4), 2298–2315 (2008)
3. Tzouveli, P., Mylonas, P., Kollias, S.: An intelligent e-learning system based on learner profiling and learning resources adaptation. Computers & Education 51(1), 224–238 (2008)
4. Meo, P.D., Garro, A., Terracina, G., Ursino, D.: Personalizing learning programs with X-Learn, an XML-based. "user-device" adaptive multi-agent system, Information Sciences 177(8), 1729–1770 (2007)
5. Shee, D.Y., Wang, Y.-S.: Multi-criteria: evaluation of the web-based e-learning system: A methodology based on learner satisfaction and its applications. Computers & Education 50(3), 894–905 (2008)
6. Kritikou, Y., Demestichas, P., Adamopoulou, E., Demestichas, K., Theologou, M., Paradia, M.: User Profile Modeling in the context of web-based learning management systems. Journal of Network and Computer Applications, Corrected Proof (November 19, 2007) (in press)
7. Chen, C.-M., Liu, C.-Y., Chang, M.-H.: Personalized curriculum sequencing utilizing modified item response theory for web-based instruction. Expert Systems with applications 30(2), 378–396 (2006)
8. Xu, R., Wunsch II, D.: Survey of clustering algorithms. IEEE Transactions on Neural Networks 16, 645–678 (2005)
9. Fu, H., Foghlu, M.: A conceptual subspace clustering algorithm in e-learning. In: 10th International Conference on Advanced Communication Technology, ICACT 2008 - Proceedings, pp. 1983–1988 (2008)
10. Zakrzewska, D.: Cluster analysis for users' modeling in intelligent E-learning systems. In: Nguyen, N.T., Borzemski, L., Grzech, A., Ali, M. (eds.) IEA/AIE 2008. LNCS, vol. 5027, pp. 209–214. Springer, Heidelberg (2008)

Automatic Generation of Processes Based on Processes Semantic for Smartflow Pattern

Lizhen Cui and Haiyang Wang

School of Computer Science and Technology, Shandong University
Jinan, 250101, China
clz@sdu.edu.cn

Abstract. Web based business processes has many particular characteristics. Traditional workflow pattern is not applied in this environment. In this paper, smartflow pattern, a new business processes application pattern, is proposed. It enables automatic generation of business process based on user's requirements. An automatic generation method based on processes semantic, the key issue of smartflow pattern, is also provided. Finally, an application of smartflow pattern, the IPVita (Intelligent Platform of Virtual Travel Agency) is described.

1 Introduction

Many works in our life include business process, especially on the web But the facilities that we can choose to manage business process on the web are limited.

Web-based business process management has many own characteristics. Traditional methods of business process management, such as workflow architecture, are not adapted in this environment any more. For example, the users of web-based business process management are individual, not enterprise users. Towards enterprise user, business processes in enterprise are usually fixed. A type business process can be designed in advance, and then many business process instances belong to this type business process can be followed. But in web-based environment, towards to many individuals, the limited business processes types designed in advance is not adapted many unpredictable requirements of individuals. Each requirement of individual may has many its own individuations. So the pattern of traditional business process management, once designed and repetitious used, is not used in web environment towards internet users.

In order to satisfy the business process management requirement of internet individual, we propose a novel process application pattern, called smartflow pattern, in which business processes including individuations can be automatic generated according to process requirements of users. Smartflow pattern is not a new method, but an new application pattern in web environment towards to business process. Lots of methods relative web-based process management still belong to traditional process pattern[1,2].

The scope of this paper is limited to discuss an automatic generation method of business processes, which is the key problem in Smartflow pattern.

L. Chen et al. (Eds.): APWeb and WAIM 2009, LNCS 5731, pp. 132–137, 2009.

The remainder of this paper is organized as follows. First we introduce this process automatic generation method as a whole. In section 3, we explicate the process semantics repository structure. In section 4, we introduce the description of user process requirements. In section 5, we introduce the algorithm of process automatic generation in detail. In section 6 we give the application of this method in our project IPVita (Intelligent Platform of Virtual Travel Agency)[3], which is typical Smartflow pattern. Section 7 concludes this paper.

2   Method Outline

In this section, we introduce the principle and architecture of this process automatic generation method.

In this paper, we choose knowledge based method, and use ontology to describe and manage process semantics [4]. What are the process semantics? We think that the concepts and associations between concepts in business processes are important. Using it, we can model process requirements of users, and generate individual process automatically. The method outline is followed:

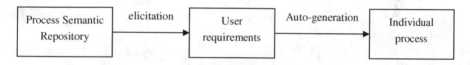

Fig. 1. Method Outline

In this architecture, the process generation fell into three phases. First, a process semantic repository is constructed. Using it, process requirements of individuals can be elicited. Finally, the process generation method can help to form a business process automatically.

3   Process Semantics Repository

We want to manage and use process semantics, such as descriptions of activities and associations between activities. Ontology technology is the best choice to present it.

3.1   Definition of Process Ontology

Definition 1. Process ontology is composed of process concepts and associations. Process concepts mainly include descriptions of activity and event. Associations mainly include the activity flow model.

We describe these definitions respectively.

3.2   Process Concepts

```
Activity:=UnitActivity|ComActivity
ComActivity:=SeqActivity|OrActivity|XorActivity
UnitActivity=(trigger:event;in:state;out:state)
```

```
OrActivity|XorActivity=(SubActivity:Activity; trig-
ger:event;in:state;out:state)
```

```
SeqActivity =(SeqActivity:Activity; trig-
ger:event;in:state;out:state)
```

Event concept includes request event, change event and time event.

```
Event:=RequestEvent|TimeEvent|ChangeEvent
```

3.3 Association Semantic

Activity flow model present the associations between activities and process in organization.

```
ActivityFlowModel=(Node, Link).
Node=(Event∪State∪Activity∪Connector)
```

$$connector=\{\vee, \wedge, \neg, =\}$$

```
Link=((Event*Activity) ∪ (State*Event) ∪ (State*Activity) ∪ (Ac
tivity*State) ∪ (State*connector) ∪ (connector*State)
∪ (Activity*connector) ∪ (connector*Activity) ∪ (connector*con
nector))
```

4 Process Requirement Elicitation

Process ontology can guide the user to form the process requirement [5]. Each process requirement is one instance of the process ontology.

Formally, we give the description of process requirement.

Definition 2. Process Requirement=(DomainCon, ConMapping, Ass).

DomainCon indicates domain concepts given by user. ConMapping indicates the relation between concepts in process ontology and it. Ass indicates the association between DomainCons.

Tab1. shows the part of process requirement in travel domain.

Table 1(a). Concepts of process requirement in travel domain

Domain Concept	Concept Mapping
Plan	Activity
Apply(visa)	Activity
Order(hotel)	Activity
Order(ticket)	Activity
Price	Data
Source	Data
Destination	Data
Ticket	Data
...	...

Table 2(b). Associations of process requirement in travel domain

Source Concept	Target Concept	Association Name
Request(travel)	Plan	Trigger
Plan	Order(ticket)	Subactivity
Plan	Apply(visa)	Subactivity
Plan	Order(hotel)	Subactivity
Time	Order(ticket)	In
Price	Order(ticket)	In
Source	Order(ticket)	In
Destination	Order(ticket)	In
Order(ticket)	Ticket	out
...	...	...

5 Process Automatic Generation Algorithm

As shown in above, activity flow model is composed of four type nodes. We can start with unitactivity, construct activity connection, and finally generate the activity flow.

Algorithm 1. Process Automatic Generation Algorithm

1. create one node for unitactivity, then draw a trigger arc to this node, the source of trigger arc is the trigger event of this unitactivity, called Trigger(A), a state input arc to this node, and a state output arc from this node. The source of state input arc is link the in state of this activity, and target of state output arc is link the out state of this activity.
2. analysis state input arc
   - if in state is null, then delete it
   - if in state is atomic, then replace it
   - if in state is composed one, replace it according to Fig.2(a) all these in states are called InStates(A)
3. analysis state output arc
   - if out state is null, then delete it

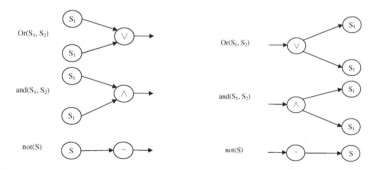

Fig. 2(a), Fig. 2(b). Process Automatic Generation Algorithm Rules

Fig. 2(c). Process Automatic Generation Algorithm Rules

- if out state is atomic, then replace it
- if out state is composed one, replace it according to Fig.2(b)
  all these out states are called OutStates(A)
4. $A_1, A_2 \in$ ActivitySet, if $s_1 \in$ InState(A_1), $s_2 \in$ OutState(A_2) and $s_1 = s_2$, connect A_1 and A_2 according to Fig.2(c)

6 Application in IPVita Project

IPVita (Intelligent Platform of Virtual Travel Agency) is a platform servicing for travel domain. It is a typical Smartflow Pattern. Each user just provides his own requirement relative his travel, and then a travel plan is automatically generated. Platform can also execute this travel process through interchanging with user.

Fig.3 is one result generated using this method in this platform.

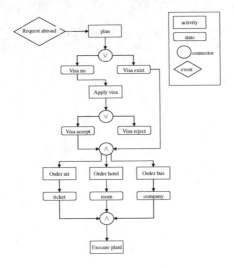

Fig. 3. A Case Study

7 Conclusion

This paper starts with the discussion that traditional process application pattern is limit in web-base process management. So a novel process application pattern, smarflow pattern, is proposed. Few researches have been found on this field.

We mainly introduce the key technology in smartflow pattern, namely, process automatic generation method. It includes process ontology construction, process requirement elicitation and process automatic generation. In addition, the application, IPVita, and a case study are introduced to illustrate this method.

In the future, we plan to provide analysis method to verify the generated process model.

References

1. Patil, A., et al.: METEOR-S web service annotation framework, New York, NY, United States (2004)
2. Gou, H., et al.: Agent-based approach for workflow management, Nashville,TN,USA (2000)
3. Sui, Q., Wang, H.-y.: IPVita: An intelligent platform of virtual travel agency. In: Zhou, X., Li, J., Shen, H.T., Kitsuregawa, M., Zhang, Y. (eds.) APWeb 2006. LNCS, vol. 3841, pp. 1205–1208. Springer, Heidelberg (2006)
4. Lizhen, C.: Research and implement on business process driven dynamic application integration over internet (Ph.D.dissertation). Shandong University (2005)
5. Jin, Z., Bell, D.A., Wilkie, F.G.: Requirements elicitation based on enterprise ontology and domain ontology. In: Proceedings of the 4th World Multi-Conference on Systems. Cybernetics and Informatics. Orlando,Florida, USA, pp. 328–333 (2000)

Easy Flow: New Generation Business Process Model

Qi Sui[*], Dong-qing Yang[*], and Teng-jiao Wang[*]

School of Electronics Engineering and Computer Science, Peking University, China
{suiqi,dqyang,tjwang}@pku.edu.cn

Abstract. Now workflow technique is almost the unique candidate to handle a business process in a software application. But graph based traditional workflow is not enough flexible in software engineer which is well known since it was introduced from manufacture industry in 1980s. Until now how to make workflow more flexible is also an unresolved problem in workflow technique which is a serious bottleneck for its popularization, although many efforts have been done to advance it. Author thinks it is wrong to design a process model based on graph theory, which is just the reason to cause workflow so inflexible. In this paper, author presents new generation process model – Easy Flow based on task and its semantic, which is easy to understand, design, modify and refactor a business process. Easy Flow model is completely different with workflow model, which is more adept to control a "changing" business process in an enterprise application. Easy Flow will replace workflow to realize real business process management in future applications.

Keywords: Business Process, Process Refactoring, Workflow, Graph.

1 Introduction

How to handle business process in an enterprise application is a common problem in software engineer. Now these works are completed by workflow or workflow based BPM techniques in most solutions, which are based on petri-net or other graph based theory. Now graph based workflow technique has almost been the unique candidate to handle a business process in a software application. In most enterprise applications solutions, workflow middleware is used to control business process and business classes are used to handle business logics.

But there are many problems when we integrate graph based workflow middleware into an enterprises application, which have been discussed many years since workflow was introduced from manufacture industry in 1980s.

The biggest problem is that workflow can't handle "changing processes". How to change a running business process is a difficult problem which has been discussed in recent 20 years. If a process is changed in running state, you must precisely analyze its impacts for all started process instance and then handle these impacts to avoid any

---

[*] Dr. Qi Sui, his research areas are workflow, BPM and their applications. Prof. Dong-qing Yang, her research areas are database technical and data mining. Dr. Teng-jiao Wang, his research areas are database technical and data mining.

L. Chen et al. (Eds.): APWeb and WAIM 2009, LNCS 5731, pp. 138–147, 2009.

risks. It is a hard work in a graph based process model with various running instance, which would cause many unresolved difficult problems including some NP problems in graph theory. Therefore a safe solution for "changing process" is that old instance use old process model and new instance for new process model, in another word, process changing is valid only for those new instances.

But on the other hand, a "changing process" is eagerly needed by most enterprise. Current world is a changing world, so an enterprise must continually change itself to adept to the new environment if it wants live longer. For example of Chinese apartment loan, recently central bank publics new loan discount for the first apartment of a consumer to avoid house market crisis. It is very suddenly and banks must add a subprocess of new discount applying in their loan process as soon as possible, otherwise consumer would leave to other bank if they can't get new discount quickly. At the same time some old process instance would have run many years for example a bank loan handling instance, so the current solution -- "new process for new instance" -- is not satisfied for enterprise of course. In some awkward cases, enterprises even ask developers to remove or round workflow solution temporarily to satisfy their urgent "process changing" requirements.

Another problem is that workflow can't handle "unclear processes". In current workflow solution, designer must define all details of current process and prearrange some nodes for future process. But in this solution, predefining for future process is almost an impossible completed task for designer because they don't know what will happen in the future. Also for example of apartment loan, many banks are trying to use more flexible payment pattern -- customer decide the monthly payment amount in a range -- to fetch more customers, but this must get warrant from central bank. Although how to modify payment hasn't been decided, it doesn't affect loan applying. Designer can only define loan applying process to run, and then add payment process in the future when payment modification is approved.

"Unclear processes" is also very common in most enterprise because we should pay more attention to complete current work than to forecast future works. The current solution -- "design for everything both now and future" -- cost too much times for tasks that maybe not happen for ever. In some worst cases, enterprise users even complain that they pay so much money to get some garbage never used.

In general workflow technique is not enough flexible for enterprise application, and it is more adept to stable business processes not to "changing" and "unclear" ones. Until now how to make workflow more flexible is also an unresolved problem which is a serious bottleneck for its popularization, although many efforts have been done to advance it. After achieving many process projects, author thinks maybe it is wrong to define a process model based on graph theory, and graph structure is too complex for business process to cause it so inflexible. In this paper, author presents new generation process model – Easy Flow based on a simple task basket, which is easy to understand, design, modify and refactor a business process. Easy Flow model is completely different with workflow model, which is more adept to control business process in an unstable enterprise environment. We believe that Easy Flow will replace workflow to realize perfect business process management in future applications.

The remainder of this paper is organized as follows. In Sec.2 some research works about how to advance current workflow techniques are discussed. In Sec. 3, new process model – Easy Flow is presented with an actual case about changing process, and it is easy to change an unclear process without any impacts for all running instances. And in last section, there are conclusions.

2 Related Works

Workflow techniques[1] came from manufacture industry in 1980s, which was first used to handle office documents automatically. In the early 1980s, there are many software including WorkFlo[2] and ViewStar[3] using different workflow models. After W. M. P. van der Aalst[4] and PNabil R. Adam etc[5] introduced petri-net as the concept model of workflow, petri-net have been accepted by most researchers. Now most workflow definition languages such as jPDL[6] and BPEL[7] are based on perti-net model to control task process. Although there are also some other workflow model at that time, all of them is still based on graph theory because most researcher and developers think a workflow model is a task handling graph. In the early of 2000s, petri-net based workflow technique was used in business process management system[8] to replace rule based techniques[9]. Now workflow technique is almost the unique candidate to handle a business process in a software application.

But both researchers and developers have found that graph based workflow model is not enough flexible for business process. The biggest problem is that workflow can't handle "changing processes" and "unclear process", but it is just what enterprise want workflow do. Once a workflow model is built to be run, changing it is a very hard work. Therefore how to make workflow model more flexible become a hot topic in workflow research. In the early of 1990s, Marc Voorhoeve etc[10] provided Ad-hoc Workflow solutions to enhance the flexibility of workflow, and R. Siebert[11] presented an open architecture for adaptive workflow management systems. Using exception handling to ensure a safe workflow changing is another efforts in 2000s, Claus Hagen etc[12] introduced a exception handling algorithm to evaluate and control exceptions in workflow changing. Wil M. P. van der Aalst etc[13] used another method -- case handling – to try to simplify the workflow changing. At the same time more basic effort had become in Alessandra Agostini etc[14], who try to simplify process model using a more simple model – Simple Process. More researches can be found in Wil M. P. van der Aalst's survey[15]. Although many perfect works have published in recent 20 years, there are no any papers can resolve all problems in workflow changing, and all workflow software have to use an old but safe policy – new model for new instances and old model for old instances. When BPM (Business Process Management) is popular after 2000, this problem becomes more serious because changing is a basic feature in business process. Jussi Stader[16] and Daniela Grigori[17] introduced how to make a BPM system more intelligent. In fact most papers aimed to make a business process more flexible in BPM research area but there are no good ideas to be used in current BPM software until now. We should think about some questions again: Why process changing is so common in our life but current graph based workflow techniques are so difficult to complete this work? Whether there are some things wrong in current workflow theory? Whether graph structure is too complex to describe a business process and whether a business process just is a task or business graph? Is it important in a business process to know which its previous tasks are when we do a task?

Some researchers have noticed this problem and they try to use a different but simple model to describe business process. Keith Harrison-Broninski[18] presented HIM(Human Interactions Management) to model a human interaction process, which is focus on human interaction rather than task process. Prashant Doshi etc[19] used Markov Decision Processes to describe dynamic workflow. Author's another paper[20] introduced a new process model – Smart Process, which is adept to complex services oriented environments. In recent years, GTD[21] (Getting Things Done) becomes very popular to introduce a simple idea to handle our tasks in our real life – put tasks into our basket with some rules and then do them.

We get ideas from these researches especially in HIM and GTD, and we provide a plain idea to describe business model. In this paper, author presents a new business process model to replace current graph based workflow model, which is based on a plain task basket rather than current complex task handling graph. This idea is plain and easy to understand, design, modify and refactor a business process.

3 Easy Flow

In this section we will present a practice case about business process changing first. And then we provide our Easy Flow model using more simple and plain concepts. At last we show how easy to change a running business process without any stopping.

3.1 A Case of Changing Process

In China there is a special social insurance named Government Apartment Fund. After you paid for it some months you can get a preferential apartment loan with lower rate and lower first pay rate. How to apply and pay a government apartment loan is more complex than common bank apartment loan, as well as it always changes to be used a balance tool by government to adjust China apartment market.

Figure 1 is a process about how to get and pay back a government apartment loan, whose detail steps are like following:

Step 1: User submits apply documents to apartment fund bureau.
Step2: Employer of apartment fund bureau checks whether apply documents are all ready, and it also named "First Check".
Step 3: Following steps are parallel done, and they are also named "Second Check".
　　Step 3.1: Employer of apartment fund bureau computes loan total amount and first pay money.
　　Step 3.2: Employer of apartment fund bureau checks loan history to judge whether you can apply.
　　Step 3.3: Employer of apartment fund bureau check paid fee history to judge whether you can apply.
　　Step 3.4: Employer of apartment fund bureau estimates the worth of your apartment.
　　Step 3.5: Employer of apartment fund bureau checks loan duration.
Step 4: Employer of apartment fund bureau computes credit score if you want get more loans. This step is an optional one.
Step 5: Sign loan contract.
Step 6: If user wants, user can apply to modify the monthly payment amount. This step is an optional one.

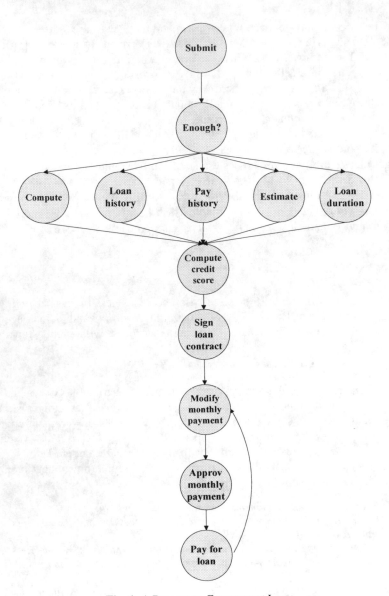

Fig. 1. A Process to Government Loan

Step 7: Employer of apartment bureau approves the modification of the monthly payment amount.
Step 8: User pays for loan every month.
Steps 6-8 are circle executed.

But in these days, this process is changing because of the OA technical developing and the recent apartment market crisis of China. Especially in this year, this process changed more and more frequent to excite China apartment market not to crash down. Different changes have happened like following:

a) After using OA technical, some check steps (Step2, Step3.2, Step3.3) can be composed into Step1 in the future, and this new step can be done by computer automatically.

b) Now some steps (Step3.3, Step3.4) had been removed for special apartment supporting of low income citizen.

c) In most cities of China, some steps (Step6, Step7) are forbidden that a customer can't change monthly payment amount.

But in some big cities, these steps are done with strict control that customer can change monthly payment amount no more than 1 or 2 times in whole loan duration. And these steps are responsible for specific employers.

In Beijing, these steps can be done freely every month because these works is done by a flexible Loan Payment System, and this pattern is applauded by most citizens. Beijing pattern is a good example for all Chinese cities, most cities will adjust their process for loan payment in recent years.

d) For different check steps (Step2, Step3.1, Step3.2, Step3.3, Step3.4, Step3.5), some different employers have different responsibility to do specific checks. But these responsibilities are changing very often with leave of old employers and coming of new employers.

e) In the future maybe some new check steps will be added to prevent cheating loan, and some new credit compute steps will be added to control loan risk.

3.2 Easy Flow

In our many years workflow experiences, we found that current graph based workflow theory is not flexible for common "changing process" and "unclear process", which give us some unsuccessful stories. We have thought about some questions in these years:

Why process changing is so common in our life but current graph based workflow techniques are so difficult to complete this work? Whether there are some things wrong in current workflow theory? Whether graph structure is too complex to describe a business process and whether a business process just is a task or business graph? Is it important in a business process to know which its previous tasks are when we do a task? What is a real business in our life, ordered task nodes or some other more simple things?

We consider again and again about what is a real process in our life, and then we get some conclusions: Business process is not a series ordered task nodes, and it is not important which its previous tasks are when we do a task.

It is real important in a business process that:

Which tasks we should complete?

When we do these tasks?

Who will do these tasks?

What will be done?

How to do these tasks?

We think if a process model can answer these questions, it is a good one, and how to change it is not a problem again. Therefore a new process model – Easy Flow is provided which is composed by some important concepts:

Project: A goal that has multiple actions associated to it in order to reach completion.
Context: A container with any constants and variables when a project is being finished.
Action: Task will be perfromed manually or automatically.
An action has some properties to answer some important question about a task:
   Invoke Condition and Priority: When we do this task?

User, Role and Authorization: Who will do this task?
Input: What will be done in this task?
Output: How to do this task?

Using Easy Flow definition, a business process model mentioned in subsection 3.1 is defined like following:

Action 1: Submit apply documents
Condition: Null
PRI: Medium
User: Any appliers
Auth: Any one
Input: Apply documents
Output: Submitting is finished

Action 2: Whether apply documents is enough
Condition: Submitting is finished
PRI: Medium
User: Employer 1
Auth: Basic Employer
Input: Apply documents
Output: Apply documents are enough or not

Action 3.1: Compute loan total amount and the first pay money
Condition: Apply documents are enough
PRI: Medium
User: Employer 2
Auth: Advanced Employer
Input: Apply documents
Output: Loan total amount and the first pay money

Action 3.2: Check loan history to judge whether you can apply
Condition: Apply documents are enough
PRI: High
User: Employer 3
Auth: Advanced Employer
Input: Apply documents
Output: Loan history is pass by the check or not

Action 3.3: Check paid fee history to judge whether you can apply
Condition: Apply documents are enough
PRI: Medium High
User: Employer 4
Auth: Advanced Employer
Input: Apply documents
Output: Paid fee history is pass by the check or not

Action 3.4: Estimate the worth of your apartment
Condition: Apply documents are enough
PRI: High
User: Employer 5
Auth: Expert Employer
Input: Apply documents
Output: The worth of your apartment and whether it is worth for this loan

Action 3.5: Check loan duration
Condition: Apply documents are enough
PRI: Low
User: Employer 6
Auth: Basic Employer
Input: Apply documents
Output: The loan duration is OK or not

Action 4: Compute credit score for more loan
Condition: First pay, Loan history, Paid fee history, Worth of your apartment, Loan duration are OK all, but Loan total amount is too much
PRI: High
User: Employer 7
Auth: Expert Employer
Input: Apply documents
Output: Credit score and new Loan total amount

Action 5: Sign loan contract
Condition: First pay, Loan history, Paid fee history, Worth of your apartment, Loan duration, Loan total amount are OK all
PRI: High
User: Employer 8
Auth: Basic Employer
Input: Apply documents
Output: Signed loan contract and mortagage contract or no-signed contract

Action 6: Apply for modification of the monthly payment amount
Condition: Signed loan contract and every month
PRI: Medium High
User: Application 1
Auth: Automatic Application
Input: ID and New monthly payment amount
Output: Apply is accepted or not after first check

Action 7: Approve for modification of the monthly payment amount
Condition: Apply is accepted
PRI: Medium High
User: Application 1
Auth: Automatic Application
Input: ID and New monthly payment amount
Output: New monthly payment is accepted

Action 8: Pay for loan every month
Condition: Every month
PRI: High
User: All appliers
Auth: Automatic Application
Input: ID and monthly payment
Output: Loan of this month is full paid or no full paid remind

Fig. 2. A Process Definition for Government Loan Using Easy Flow

3.3   Process Changing Using Easy Flow

In Easy Flow we pay more attentions on task actions and their semantic properties, not on order of task nodes in traditional workflow, and task order is describe with action conditions. So it is easy to change an action in a project than a node connected with many other ones in a workflow graph.

Because task order is not important in Easy Flow, a project can be changed easily without any impact for running instances. In a running instance any actions haven't been done can be perfrom using new action definition. It is a big difference with traditional workflow, in which any running instances can only using old process definition until it is finished. Changing for a business process can be done in any time without any stop in process running.

Using Easy Flow definition, a process changing is very easy without any impact on running instance like following:

a) Some check steps (2, 3.2, 3.3, 3.5) can be composed into step 1.

Modification: Remove action 2, action 3.2, action 3.3, Modify action 1 as following:
Action 1: Submit apply documents and first check
  Condition: Null
  PRI: Medium
  User: Application 1
  Auth: Automatic Application
  Input: Apply documents
  Output: Apply documents is enough, Loan history, Paid fee history, Loan duration are OK all

b) Some steps (3.3, 3.4) had been removed for special apartment.

Modify conditions of action 3.3, action 3.4 as following:
Action 3.3: Check paid fee history to judge whether you can apply
  Condition: Apply documents are enough and applier's income is not low
  PRI: Medium High
  User: Employer 4
  Auth: Advanced Employer
  Input: Apply documents
  Output: Paid fee history is pass by the check or not
Action 3.4: Estimate the worth of your apartment
  Condition: Apply documents are enough and applier's income is not low
  PRI: High
  User: Employer 5
  Auth: Expert Employer
  Input: Apply documents
  Output: The worth of your apartment and whether it is worth for this loan

c) Some steps (6, 7) are not forbidden in all cities.

Modify action 6 as following:
Action 6: Apply for modification of the monthly payment amount
  Condition: Signed loan contract and modification times <1
  PRI: Medium High
  User: Application 1
  Auth: Automatic Application
  Input: ID and New monthly payment amount
  Output: Apply is accepted or not after first check, and if accepted modification times ++

d) Responsibilities are changing very often.

Modify users in action 2, 3.1, 3.2, 3.3, 3.4, 3.5 as following:
Action 6: Apply for modification of the monthly payment amount
  Condition: Signed loan contract and modification times <1
  PRI: Medium High
  User: Employer new 1
  Auth: Advanced employer
  Input: ID and New monthly payment amount
  Output: Apply is accepted or not after first check, and if accepted modification times ++

e) Some new check steps will be added to prevent cheating loan.

Add new actions easily:
Action 3.6: Check cheating marriage.
  Condition: Apply documents are enough
  PRI: High
  User: Police Officer
  Auth: External user
  Input: Apply documents
  Output: It is a true marriage or not

Fig. 3. New Process Definition for Government Loan Using Easy Flow

4 Conclusions

There are some problems in traditional graph based workflow model since workflow was introduced from manufacture industry in 1980s. Workflow can't handle "changing processes" and "unclear processes", but they are very common in an enterprise application. In general workflow technique is not enough flexible in software engineer, and it is more adept to stable business processes not to "changing" and "unclear" ones. Until now how to make workflow more flexible is also an unresolved problem for workflow technique which is a serious bottleneck for its popularization, although many efforts have been done to advance it.

Author of this paper thinks maybe it is wrong to define a process model based on graph theory, and graph structure is too complex for business process to cause it so inflexible. In this paper, author presents new generation process model – Easy Flow based on a simple task basket, which is easy to understand, design, modify and refactor a business process. Easy Flow model is completely different with workflow model, which is more adept to control business process in a "changing" and "unclear" enterprise environment. We believe that Easy Flow will replace workflow to realize perfect business process management in the future applications.

Acknowledgments. This work is supported by the Cultivation Fund of the Key Scientific and Technical Innovation ProjectMinistry of Education of China(No.708001), the National '863' High-Tech Program of China(No.2007AA01Z191,2006AA01Z230), and the NSFC(Grants 60873062).

References

1. Van Der Aalst, W.M.P., Ter Hofstede, A.H.M., Kiepuszewski, B., Barros, A.P.: Workflow Patterns. Distributed and Parallel Databases 14(1) (July 2003)
2. McGuire, R.S.: Implementing a filenet WorkFlo system for personnel records management, Case studies of optical storage applications (October 1990)
3. Moad, J.: ViewStar faces many Goliaths. Datamation 38(11) (May 1992)
4. van der Aalst, W.M.P.: Woflan: a Petri-net-based workflow analyzer. Systems Analysis Modelling Simulation 35(3) (May 1999)
5. Adam, N.R., Atluri, V., Huang, W.-K.: Modeling and Analysis of Workflows Using Petri Nets. Journal of Intelligent Infromation Systems 10(2) (March 1998)
6. jBPM Process Definition Language(JPDL),
   http://docs.jboss.org/jbpm/v3/userguide/jpdl.html
7. Business Process Execution Language for Web Services version 1.1, http://www-128.ibm.com/developerworks/library/specification/ws-bpel
8. van der Aalst, W.M.P., van Hee, K.M.: Business process redesign: a Petri-net-based approach. Computers in Industry 29(1-2) (July 1996)
9. Halle, B.V., Ross, R.G.: Business Rules Applied: Building Better Systems Using the Business Rules Approach. John Wiley & Sons, Inc., Chichester (2001)
10. Voorhoeve, M., van der Aalst, W.M.P.: Ad-hoc Workflow: Problems and Solutions. In: Proceedings of Eighth International Workshop on Database and Expert Systems Applications (September 1997)
11. Siebert, R.: An Open Architecture For Adaptive Workflow Management Systems. Journal of Integrated Design & Process Science 3(3) (August 1999)
12. Hagen, C., Alonso, G.: Exception Handling in Workflow Management Systems. IEEE Transactions on Software Engineering 26(10) (October 2000)
13. van der Aalst, W.M.P., Weske, M.: Case handling: a new paradigm for business process support. Data & Knowledge Engineering 53(2) (May 2005)
14. Agostini, A., de Michelis, G.: A Light Workflow Management System Using Simple Process Models. Computer Supported Cooperative Work 9(3-4) (August 2000)
15. van der Aalst, W.M.P., Basten, T., Verbeek, H.M.W., Verkoulen, P.A.C., Voorhoeve, M.: Adaptive workflow. Enterprise infromation systems (June 2000)
16. Stader, J., Moore, J., Macintosh, A., Chung, P., McBriar, I., Ravindranathan, M.: Intelligent software support for business process change. Systems engineering for business process change: new directions (January 2002)
17. Grigori, D., Casati, F., Castellanos, M., Dayal, U., Sayal, M., Shan, M.-C.: Business process intelligence. Computers in Industry 53(3) (April 2004)
18. Harrison-Broninski, K.: Human Interactions: The Heart And Soul Of Business Process Management: How People Reallly Work And How They Can Be Helped To Work Better. Meghan-Kiffer Press, Tampa (2005)

19. Doshi, P., Goodwin, R., Akkiraju, R., Verma, K.: Dynamic Workflow Composition using Markov Decision Processes. International Journal of Web Services Research (2005)
20. Sui, Q., Dongqing, Y., Haiyang, W.: Smart Process: Service Oriented Business Process Management Pattern in E-hospital. In: Proceedings of 2008 IEEE International Symposium on IT in Medicine & Education (November 2008)
21. Allen, D.: Getting Things Done: The Art of Stress-Free Productivity. Penguin Inc. (April 2001)

User Feedback for Improving Question Categorization in Web-Based Question Answering Systems

Wanpeng Song[1,2,3], Liu Wenyin[2,3], Naijie Gu[1,3], and Xiaojun Quan[2]

[1] Department of Computer Science and Technology, University of Science
and Technology of China, Hefei, China
[2] Department of Computer Science, City University of Hong Kong, Hong Kong, China
[3] Joint Research Lab of Excellence, CityU-USTC Advanced Research Institute,
Suzhou, China
wadeswp@mail.ustc.edu.cn,
{csliuwy,xiaoquan}@cityu.edu.hk,
gunj@ustc.edu.cn

Abstract. Question categorization, which automatically suggests a few categories to host a user's question, is a useful technique in Web-based question answering systems. In this paper, we propose a question categorization method which makes use of user feedback to the system's automatic suggestions to improve question categorization. We initialize the categorization model by training a set of accumulated questions. When a user asks a question, the system automatically suggests a few categories for the question using the current categorization model. The user can then select one of these suggestions or another category and such feedback information is used to revise the categorization model. The revised model is used to categorize new questions. Experimental results show that our method is effective to take the advantages of both positive and negative feedback to improve the precision of question categorization but finish the revision of the categorization model in real time.

Keywords: Question categorization, User Feedback, Web-based question answering.

1 Introduction

Web-based question answering (QA) has become a popular information service with the emergence of Web 2.0. In Web-based QA systems, such as Yahoo! Answers [15] people can help each other with human-provided answers, which overcomes the shortcoming of poor quality of automatic answers. With many people asking and answering questions every day, the systems have accumulated a large number of questions. Hence, it is necessary to organize these questions in a good way for users to browse and navigate them more conveniently. Question categorization is a technique used for this purpose, which automatically suggests a few predefined categories to host a newly posted question according to the topic or content of the question. It is very helpful for users to find interesting questions. For example, in our Web-based QA system *BuyAns* [14], when a user asks a question, the system often

L. Chen et al. (Eds.): APWeb and WAIM 2009, LNCS 5731, pp. 148–159, 2009.
© Springer-Verlag Berlin Heidelberg 2009

recommends/suggests a few (0 to 3) categories (or boards) ranked by their relevancy to the question for the user to choose. If the user finds a suitable one, he only needs to click a button to confirm it. Otherwise, he can still choose the suitable one from the entire category list. Hence, it will save much effort if the system can recommend/suggest the categories accurately.

However, it is difficult to achieve very high categorization precision initially without considering user feedback because the machine itself may not understand users' questions very well. Since user interaction is available as choosing one suitable category from the recommended list (or not) in the system, such interaction provide more information for the system to understand the question better. The user feedback can be used to revise the current categorization model for the system to categorize new questions more accurately.

In this paper, we propose a question categorization method which makes use of user feedback to improve question categorization for Web-based QA. Firstly, we initialize the categorization model by training a set of labeled questions. Each category is represented by a vector in the model. When a user asks a question, the question is automatically categorized using the current categorization model and a few categories with higher relevancy are recommended to the user to choose. The user can then interact with the system in either of the two ways: confirming one of the recommended categories or choosing one category from the entire category list. The question can be considered as a positive feedback to the selected category and a negative feedback to all other unselected categories. Such user feedback is captured and used to revise the categorization model. For each category, we use both positive and negative feedbacks to update the category vectors. For each category vector, we increase the weight of features that appear in positive questions and decrease the weight of features that appear in negative questions. The revised model is then used to categorize new questions. Experimental results show that our method is effective to take advantages of both positive and negative feedback to improve the precision of question categorization but finish the revision of the categorization model in real time.

The rest of the paper is organized as follows. In Section 2, we briefly review some related work. Initialization of our categorization model is presented in Section 3. In Section 4, we describe our method of revising the categorization model based on user feedback in details. Experimental results are shown in Section 5 and finally Section 6 concludes this paper.

2 Related Work

There have been some research works [1, 2, 3] on question type classification, which classifies questions into some semantic classes that impose constraints on potential answers, like "person", "location", "definition" and so on. In question type classification, the interrogatives such as "who" and "when" are very important because they directly indicate the question types. However, in question topic categorization for a user-interactive QA system, the situation is different, in which the categories are predefined according to the question topic, such as "computer", "education", "sports" and so on. For example, the two questions "who invented computer?" and "when did human invent computer" belong to the same category "computer" in question topic categorization, but may belong to different categories in question type classification. Hence, the content of the question plays an important role in question topic categorization.

Question topic categorization can be considered as a particular type of (short) text categorization problem, which has been investigated in many literatures. Most of (short) text categorization approaches employ machine learning techniques to construct a classifier and use it to categorize new documents, such as Naïve Bayesian classifier [4], Rocchio algorithm [5], and support vector machine (SVM) [6]. Most of these classification techniques are based on some underlying models or assumptions. For instances, SVM is based on the Structural Risk Minimization principle and Naïve Bayesian classifier is based on the assumption that the probabilities of words are independent. When the data fit the model or satisfy the assumption, the performance is usually good; otherwise, it may be quite poor.

Relevance feedback [13] has been proposed as a technique to narrow the semantic gap between machines and human beings by incorporating user feedback into machine learning processes. Many research studies have shown that relevance feedback is a useful technique for improving information retrieval performance. Raghavan et al. [8] extend the traditional active learning framework to include feedback on features in addition to labeling instances, and conduct a careful study of the effects of feature selection and human feedback on features in the setting of text categorization. Luan et al. [9] incorporate the use of relevance feedback into video retrieval. They segregate the process of relevance feedback into two distinct facets: recall-directed feedback and precision-directed feedback. The recall-directed facet employs general features such as text and high level features (HLFs) to maximize efficiency and recall during feedback, making it very suitable for large corpora. The precision-directed facet uses many other multimodal features in an active learning environment to improve accuracy. Huang et al. [10] introduce a new approach to mixed-initiative clustering that handles several natural types of user feedback for text clustering. They introduce a new probabilistic generative model for text clustering and describe how to incorporate four distinct types of user feedback into the clustering algorithm. To the best of our known, there have been few literatures on using user feedback to enhance question categorization, especially in the scenario of Web-based QA. In this paper, we make use of user feedback to revise our question categorization model in real time, which takes the advantage of human-machine interactions.

3   Initializing the Categorization Model

We initialize our categorization model by training a set of accumulated questions, which have been classified into corresponding categories in our system. Our method is based on vector space model. First, each category is represented by a prototype vector obtained from training. Second, each new posted question is also represented by a vector and the relevancy or similarity between the question and each category is calculated based on their vectors. In our model, we use cosine similarity to calculate the similarity. Using this model, the process of learning is very fast for online process and accordingly system can response quickly to users.

3.1 Feature Selection

Feature selection is to select the important terms extracted from accumulated questions to construct a feature space. Considering a question contains a few words, we use word-level features. Before feature selection, stop words are removed and the remaining words are stemmed to their roots. Information gain is chosen as the selection criteria, which has been proved to be quite effective in previous research [7]. The higher information gain a feature gets, the more strongly the feature indicates the presence or absence of a category. The information gain of a term t is defined as follows:

$$IG(t) = -\sum P(c_i) \log P(c_i) + P(t) \sum P(c_i \mid t) \log P(c_i \mid t) + P(\bar{t}) \sum P(c_i \mid \bar{t}) \log P(c_i \mid \bar{t}) \qquad (1)$$

where $IG(t)$ is the information gain of feature t, $P(c_i)$ is the probability of category c_i, which is equal to the number of questions in c_i divided by the number of questions in all categories; $P(t)$ is the probability that term t appears in a question, which is equal to the number of questions containing t divided by the number of all questions; $P(\bar{t})$ is the probability that term t does not appear in a question; $P(c_i \mid t)$ is the probability of category c_i given that term t appears and $P(c_i \mid \bar{t})$ is the probability of category c_i given that term t does not appear.

All the extracted words are sorted by their information gain scores defined in Eq. (1) in the descending order and the top k words are selected as features.

3.2 Weighting Category Vectors

The vector of each category consists of the weights of the selected features for the category. We consider the impact of the features in a specific category and the whole corpus. We define Term Frequency (TF) of a feature in a specific category as its occurrence in this category and, the Category Frequency (CF) of a feature as the number of categories in which it occurs. Each category is represented by a vector as follows:

$$V_{C_i} = (wc_{i1}, wc_{i2}, ..., wc_{ij}, ..., wc_{in}) \qquad (2)$$

Where V_{Ci} is the vector representation of category C_i in the feature space, n is the number of features and wc_{ij} is the weight of the j-th feature in category C_i, which is defined as follows:

$$wc_{ij} = TF_{ij} (\log \frac{M}{CF_j} + 1) \qquad (3)$$

Where M is the total number of categories in the corpus, TF_{ij} is the Term Frequency of the j-th feature in category C_i and CF_j is the Category Frequency of the j-th feature. If we consider a category as a document, our model is exactly the traditional TF-IDF model for text retrieval.

4 Revising the Categorization Model Based on User Feedback

User feedback is important to Web-based question categorization since human can understand questions better and user feedback can be used to improve the categorization precision. We use user feedback to revise our categorization model in term of revising the weight of each category vector. For each new question, we first use the current categorization model to categorize it by recommending at most 3 categories with the highest similarities to the new question. The results are returned to the user for judgment as feedback. The judgment/feedback is represented as either confirming one of the recommend categories or choosing a correct category from the entire category list. The user judgment is then used to adjust the weight of features in category vectors. The revised model is then used to categorize new questions. The workflow is shown in Figure 1.

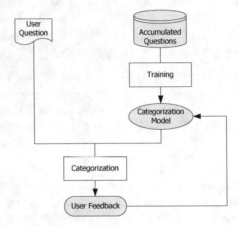

Fig. 1. Workflow of categorization with user feedback

4.1 User Question Representation

In order to measure the similarity between a question and categories, we also need to represent each user question by a vector. Question representation is somewhat different from category representation. Since a question consists of only a few words, the vector will be very sparse if we use a simple term frequency method. The basic idea of our method is to enrich the question vector by mapping the question into the feature space based on WordNet [12] and all categories.

Let q denote a question, the vector V_q of q can be represented as follows:

$$V_q = (wq_1, wq_2, ..., wq_i, ..., wq_m) \qquad (4)$$

where m is the number of words in q and wq_i is the weight of the i-th word in question q, which is equal to the term frequency of the word.

Then we convert V_q to a new vector $V_q{'}$ by mapping V_q into the feature space:

$$V_q' = (wq_1', wq_2', ..., wq_j', ..., wq_n') \tag{5}$$

Where n is number of features and wq_j' is the new weight, which is defined as follows:

$$wq_j' = \max\{wq_i s_{ij} \mid 1 \le i \le m\} \tag{6}$$

Where s_{ij} is the semantic similarity between the i-th word in q and the j-th word in the feature space, which is calculated using the path length method [11] based on Word-Net [12] by the following formula:

$$s_{ij} = \frac{1}{Dis_{ij} + 1} \tag{7}$$

where Dis_{ij} is minimum path length between the two compared words in WordNet.

After such mapping, the similarity between the question and each category can be calculated using the cosine similarity:

$$sim(q, c_i) = \frac{V_q' \cdot V_{C_i}}{\| V_q' \| \cdot \| V_{C_i} \|} \tag{8}$$

where $sim(q, c_i)$ is the similarity between the question and the i-th category, V_{Ci} is the vector of the C_i All the categories are sorted by the similarity sore in the descending order and several categories with the highest similarity scores are suggested to the user.

4.2 User Feedback

In our Web-based QA system, a user can provide his feedback with either of the two following operations according to the categories recommended by the system: clicking a button to confirm one of the recommended categories if it is correct or choosing a category from the entire category list. In such scenario, the question is a positive feedback to the selected category and a negative feedback to all the other unselected categories. Accumulated in this way, the system can capture much feedback information for each category: a set of positive questions and a set of negative questions. The two sets of questions are good resources to be used to adjust the weight of features in each category vector.

4.3 Adjusting Weight of Features in Category Vectors

When updating the category vectors, we should consider both positive feedback and negative feedback. For each category vector, we should increase the weight of features that appear in positive questions and decrease the weight of features that appear in negative questions.

Let N_i^+ be the number of positive questions of C_i. Assume that all the positive questions for a category are distributed nearby from each other in the feature space. We compute a representative vector of all positive questions as follows:

$$\overline{V_{C_i}^+} = \frac{1}{N_i^+} \sum_{j=1}^{N_i^+} V_{q_j}^+ \tag{9}$$

where $\overline{V_{C_i}^+}$ is the representative vector of all positive questions and $V_{q_j}^+$ is the vector of j-th positive question.

Let N_i^- be the number of negative questions of C_i. The distribution of all negative questions in the feature space is different from that of the positive ones since the negative question may scatter in the entire space. The reason is that the negative questions may belong to different categories while the positive questions belong to the same category. Hence, it is hard to find a suitable center for all the negative questions. However, we may still find some representative features in these negative questions. A representative feature of these negative questions should be the one which is the same or just changes a little bit among the negative questions. We argue that the weight of these features in the vectors of negative questions should vary a little around their expectations since the characteristics of features are reflected on their weights. Hence, we can use the variance to measure the variety of feature weights in these negative questions. The variance of the weights of feature f can be calculated by the following formula:

$$D(w_f) = \frac{1}{N_i^-} \sum_{j=1}^{N_i^-} \left(w_{f_j} - \frac{1}{N_i^-} \sum_{j=1}^{N_i^-} w_{f_j} \right)^2 \tag{10}$$

where $D(w_f)$ is the variance of the weight of feature f, w_{f_j} is the weight of f in the vector of j-th negative question. For each of the features extracted from negative questions, if the variance of its weight is below a threshold δ, it will be selected as one of the representative features for negative questions.

We also use a probabilistic measure to explore the degree of negative effect of a question on each category. We think that the degree of negative effect is proportional to the similarity score between the question and the category calculated by the system. The higher the similarity score is, the deeper the degree of negative effect is. This is because a higher predicting similarity value between a negative question and a category corresponds to a heavier misjudgment made by the system. Accordingly, it is necessary to decrease the impact (weight) of features appearing in both the question and the category to ensure that they will get a lower similarity score next time. Hence, we can use the similarity score to measure the degree of negative effect of a question on each category.

Let sim_{ij} be the similarity score between category C_i and the j-th negative question. We can also compute a representative vector for negative questions by the following two steps:

(1) Compute a representative vector for negative questions

$$\overline{V_{C_i}^-} = \frac{1}{N_i^-} \sum_{j=1}^{N_i^-} sim_{ij} V_{q_j}^- ;$$

(2) For each feature f in the feature space, if $D(w_f) > \delta$, then set $\overline{V_{C_i(f)}^-} = 0$, where $\overline{V_{C_i(f)}^-}$ is the weight of f in $\overline{V_{C_i}^-}$.

After both positive and negative feedback are considered, we can update the category vector using the Rocchio algorithm [5]:

$$V'_{C_i} = \alpha V_{C_i} + \beta \overline{V_{C_i}^+} - \gamma \overline{V_{C_i}^-} \tag{11}$$

where V'_{C_i} is the updated vector for C_i, α, β and γ are parameters between 0 and 1 with the constraint $\alpha + \beta + \gamma = 1$.

5 Experiment

In this section, we will first describe the data set and the evaluation metrics, and then show the experimental design and results.

5.1 Data Set

The training data in our experiment is taken from our pattern-based user-interactive QA system—*BuyAns* [14], in which all the questions are classified into 19 coarse categories and over 100 fine categories according to the question content. The number of questions in each category differs significantly from each other. Some hot topic categories like "hardware" may have more than 100 questions and some less hot ones like "religion" may have as few as 10 questions. Among all the questions accumulated in *BuyAns*, we select 7500 questions from 60 fine categories, 6000 questions for training to initialize the model, 1000 questions for simulating user feedback and 500 questions for testing. The proportion of questions for training and testing is kept almost the same in each category.

5.2 Evaluation Metrics

In Web-based QA systems, when a user asks a question, the question is automatically categorized and a few categories with higher relevancy are recommended to the user to choose. However, it is not suitable to suggest too many categories for a new question since it also costs the user's time on browsing in the long suggested list. It is more preferable to suggest only one or several highly relevant categories for the user to confirm or to choose. In our system, we suggest at most 3 categories for each new question. Hence, we use *Precision at n* (P@n) as the performance metrics, which means the proportion of questions whose correct category is within the top n categories our method suggests. For example, in our experiment, P@3=0.5

means that for half the questions, their correct categories are among the top 3 categories our system recommends.

5.3 Experimental Design and Results

To test the effectiveness of user feedback, we simulate the user feedback on the 1000 questions mentioned above. For each of the 1000 questions, we first let the system recommend suitable categories automatically and then assume its ground truth category is provided as user feedback. The 1000 questions are evenly divided into five groups. While each group of 200 questions is used as one round of feedback to revise the categorization model, the same 500 questions are used for testing the categorization precision always. The optimal parameter settings are shown in Table 1, which are obtained after extensive testing.

Figure 2 shows the categorization precision with user feedback. When the number of questions for user feedback is 0, no feedback happens and question categorization is conducted using the initial model. From Figure 2 we can see that P@1, P@2 and P@3 all increase as the number of questions used for user feedback increases. This indicates that user feedback can be made use of to revise the categorization model and gradually improve the categorization performance.

To show the effectiveness of using both positive feedback and negative feedback, we conduct experiments with five settings: (1) categorization using the initial model trained based on the 6000 training questions, (2) categorization using the initial model trained based on the 7000 training questions (the 6000 questions used for training and

Table 1. Optimal parameter settings

Parameter	Value
Number of features k	141
Threshold δ	0.01
α	0.5
β	0.3
γ	0.2

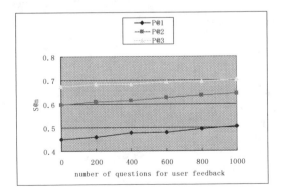

Fig. 2. Categorization performance with user feedback

the 1000 questions used for user feedback as described in Section 5.1), (3) categorization with the model revised using the 1000 questions as positive feedback only, (4) categorization with the model revised using the 1000 questions as both positive feedback and negative feedback, (5) categorization using SVM as baseline based on the 6000 training questions. Table 2 shows the comparison of these results, from which we can obtain three observations. The first observation is that the fourth setting outperforms the last. The second observation is that the third and fourth settings outperform the first setting, and the fourth setting outperforms the third setting. The third observation is that the second setting achieves the best precision among the four settings. The first observation indicates that our proposed method can achieve better performance than SVM, which is one of the best methods for classification task. The second observation indicates that not only positive feedback but also negative feedback are useful for question categorization. The third observation indicates that our method fails to achieve as good precision as retraining with equivalent questions. The reason may be that retraining can reselect the features based on some statistic information (such as information gain) while our method can only adjust the weights of the features and cannot reselect the features. However, our method still has some advantages compared to retraining, such as our method can revise the categorization model in real-time while retraining cannot because retraining costs much more time than our method. Figure 3 shows the runtime (in seconds) of retraining and our method using different number of training questions, which is tested on a 2.0GHz Dual-Core PC running Window Vista with 2 GB memory. We can see the runtime of our method varies little as the number of training questions increases and our method is much faster than retraining by about 100 times. In real use, we can make a trade off between

Table 2. Comparison of categorization accuracies of different experiment settings

Settings	P@1	P@2	P@3
(1) Initial model based on 6000 training questions	0.450	0.596	0.674
(2) Initial model based on 7000 training questions	0.528	0.660	0.714
(3) Positive feedback only	0.488	0.628	0.690
(4) Both positive feedback and negative feedback	0.504	0.644	0.702
(5) SVM based on 6000 training questions	0.498	0.632	0.692

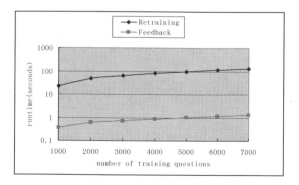

Fig. 3. Comparison of runtime of retraining and our method

efficiency and precision. For example, we retrain the data periodically (e.g., every week), and during each period, we use user feedback to revise the categorization model.

6 Conclusions

In this paper, we propose a question categorization method which can make use of user feedback to revise the categorization model. Firstly, we initialize the categorization model using a set of accumulated questions. When a user asks a question, the question is automatically categorized using the current categorization model and a few categories with higher relevancy are recommended to the user to choose. The user can then interact with the system by either confirming one of recommended categories or choosing one category from entire category list, which is captured by the system as user feedback. Such user feedback is used to revise the categorization model, which is used to categorize new questions. Experimental results show that our method is effective to take the advantages of both positive and negative feedback to improve the precision of question categorization but finish the revision of the categorization model in real time.

Acknowledgement

The work described in this paper was fully supported by a grant from City University of Hong Kong (Project No. 7002336) and the National Grand Fundamental Research 973 Program of China under Grant No.2003CB317002.

References

1. Li, X., Roth, D.: Learning Question Classifiers. In: Proceedings of the 19th International Conference on Computational Linguistics (2002)
2. Zhang, D., Lee, W.S.: Question Classification using Support Vector Machine. In: Proceedings of the 26th annual international ACM SIGIR conference on Research and development in informaion retrieval (2003)
3. Voorhees, E.: Overview of the TREC 2001 Question Answering Track. In: Proceedings of the 10th Text Retrieval Conference (TREC10), pp. 157–165. NIST, Gaithersburg (2001)
4. McCallum, A., Nigam, K.: A Comparison of Event Models for naïve Bayes text classification. In: AAAI 1998 Workshop on Learning for Text Categorization. Tech. Rep. WS-98-05. AAAI Press, Menlo Park (1998)
5. Joachims, T.: A Probabilistic Analysis of the Rochhio Algorithm with TFIDF for Text Categorization. In: Proceedings of the Fourteenth International Conference on Machine Learning (1997)
6. Joachims, T.: Text Categorization with Support Vector Machines: Learning with many Relevant Features. In: Proceedings of the Tenth European Conference on Machine Learning (1998)
7. Sebastiani, F.: Machine learning in automated text categorization. ACM Computing Surveys 34(1), 1–47 (2002)

8. Raghavan, H., Madani, O., Jones, R.: Active Learning with Feedback on Features and Instances. Journal of Machine Learning Research 7, 1655–1686 (2006)
9. Luan, H., Neo, S., Goh, H., Zhang, Y., Lin, S., Chua, T.: Segregated feedback with performance-based adaptive sampling for interactive news video retrieval. In: Proceedings of the 15th international conference on Multimedia, pp. 293–296 (2007)
10. Huang, Y., Mitchell, T.M.: Text clustering with extended user feedback. In: Proceedings of the 29th annual international ACM SIGIR conference on Research and development in information retrieval (SIGIR 2006), pp. 413–420 (2006)
11. Rada, R., Mili, H., Bichnell, E., Blettner, M.: Development and Application of Metric on Semantic Nets. IEEE Transactions on Systems, Man, and Cybernetics 9(1), 17–30 (1989)
12. Fellbaum, C.: WordNet: An Electronic Lexical Database. MIT Press, Cambridge (1998)
13. Williamson, E.R.: Does relevance feedback improve document retrieval performance? In: Proceedings of the 1st annual international ACM SIGIR conference on Information storage and retrieval (SIGIR 1978), pp. 151–170 (1978)
14. Buyans, http://www.buyans.com
15. Yahoo Answers, http://answers.yahoo.com/

XTree: A New XML Keyword Retrieval Model[#]

Cong-Rui Ji, Zhi-Hong Deng[*], Yong-Qing Xiang, Hang Yu, and Shi-Wei Tang

Key Laboratory of Machine Perception (Minister of Education), School of Electronics Engineering and Computer Science, Peking University, Beijing 100871, China
zhdeng@pku.edu.cn

Abstract. As more and more data are represented and stored by XML format, how to query XML data has become an increasingly important research issue. Keyword search is a proven user-friendly way of querying HTML documents, and it is well suited to XML trees as well. However, it is still an open problem in XML keyword retrieval that which XML nodes are meaningful and reasonable to a query, how to find these nodes effectively and efficiently. In recent years, many XML keyword retrieval models have been presented to solve the problem, such as XRANK and SLCA. These models usually return the most specific results and discard most ancestral nodes. There may not be sufficient information for users to understand the returned results easily. In this paper, we present a new XML keyword retrieval model, XTree, which can cover every keyword node and return the comprehensive result trees. For XTree model, we propose Xscan algorithm for processing keyword queries and GenerateTree for constructing results. We analytically and experimentally evaluate the performances of our algorithms, and the experiments show that our algorithms are efficient.

1 Introduction

Because XML is gradually becoming the standard in exchanging and representing data, effective and efficient methods to query XML data has become an increasingly important issue. Traditionally, research work in this area has been following one of the two paths: the structure-based query language, such as XPath [1], XQuery [2] and other fully structure-based languages [3]-[10], and the keyword-based search, such as XRANK [11], XKSearch [12] and XSeek [13].

Keyword search has been proven to be a user-friendly way of querying HTML documents, and it is well suited to XML data as well. Keyword search overcomes the problems that exist in structure-based query models, so users can get desired results without the information about schemas. However, the absence of structural information leads to the lack of expressivity, thereby sometimes users may not find the needed information.

[#] Supported by the National High-Tech Research and Development Plan of China under Grant No.2009AA01Z136.

[*] Corresponding author.

L. Chen et al. (Eds.): APWeb and WAIM 2009, LNCS 5731, pp. 160–171, 2009.

For example, assume an XML document, named "Supermarket.xml", modeled using the conventional labeled tree, as shown in Figure 1. It contains the information about a supermarket, including the locations, managers, merchandises, etc. A user interested in finding the relationship between "John" and "Bob" issues a keyword search "John Bob" and the retrieving system should return the nodes <1.4.10.10>, <1.4.10.10.2> and <1.4.10.20.2>. The meaning of the answers is easy to understand according to the document tree: The node <1.4.10.10> means that Bob is the author of a book about John; and the nodes <1.4.10.10.2> and <1.4.10.20.2> mean that Bob and John are both authors of the same books. Another example is the keyword search "John Wal-Mart", the retrieval system should return the nodes <1> and <1.4.10.20>. Node <1> means that John Lee is the manager of Wal-Mart located in Beijing, and node <1.4.10.20> means that John is the author of a book about Wal-Mart.

However, most XML keyword retrieval models, such as XRANK[11] and SLCA [12], tend to return the most specific results. For the query "John Bob", neither SLCA nor XRANK will return the node <1.4.10.10> as a result, and for query "John Wal-Mart", neither SLCA nor XRANK will return the node <1> as a result, though these nodes do contain meaningful information.

For dealing with the shortcomings of XRANK model and SLCA model, we present a new XML keyword retrieval model, Xtree, which returns more meaningful results than XRANK and SLCA. The new model has two important features. First, the results

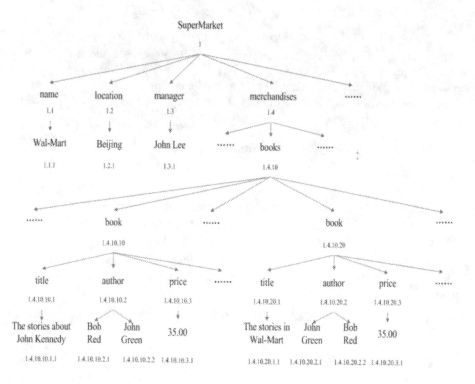

Fig. 1. Supermarket.xml (encoded by Dewey IDs)

of the XTree model will cover every keyword node, that is, a keyword node will be the descendant of at least one result node. Second, the XTree model will connect the results to corresponding keyword nodes to get result trees, which can help users easily find needed information. In addition, we also propose an algorithm of linear complexity to implement the keyword query of our model.

The paper is organized as follows. Section 2 presents the XTree model and corresponding algorithms. Section 3 discusses the experiment results. We have the conclusion in Section 4.

2 XTree: Definition and Algorithms

We present XTree model in this section. XTree model returns more results than XRANK model and SLCA model, and it still maintains a linear complexity. XTree model also returns comprehensive result trees, which can cover every keyword node.

Let $Q = \{k_1, k_2, ... k_m\}$ indicates a query contains m keywords, let $R_0 = \{v \mid \forall k \in Q(contains(v,k))\}$ indicates the nodes which contain all keywords. The predicate $contains(v,k)$ here indicates that node v contains keyword k. The results of XTree are defined as follows:

$$Result(XTree) = \{v \mid v \in R_0 \land ((subs(v) - R_0) \neq \Phi \rightarrow$$
$$\exists k \in Q \exists u \in (subs(v) - R_0)(contains(u,k)))\}$$

The predicate $subs(v)$ in the definition indicates all sub nodes of node v. The definition tells that a result of XTree must satisfy two restrictions. First, it contains all keywords. Second, after excluding the sub nodes containing all keywords, it still contains at least one keyword. According to the definition, node <1.4.10.10> is a result for query "John Bob", and node <1> is a result for query "John Wal-Mart".

2.1 Xscan: An Efficient Query Algorithm

Before presenting our query algorithm, we first describe some definitions and Lemma.

We define that the distance of two nodes is their common prefix of Dewey ID. The bigger the distance of two nodes is, the nearer the two nodes are. Let a and b are two keyword nodes, the function $LCA(a,b)$ computes the lowest common ancestor of a and b.

Lemma 1. Let c be a result of XTree model, there must be two keyword nodes a and b, where $c=LCA(a,b)$, and a is the nearest neighbour of b, or b is the nearest neighbour of a.

Proof. By the definition of XTree model, node c contains both keywords, and after excluding the sub nodes which contains both keywords, it still contains at least one keyword node, assume this keyword node is a, and b' is the nearest neighbour of a, and $c'=LCA(a,b')$, then $c'=c$, or c' contains both keywords, it will be excluded, and c will not contain a, educing a contradiction. We replace b' with b, and get $c=LCA(a,b)$, and b is the nearest neighbour of a. The similar case happens again if the keyword node contained by c is b.

Lemma 1 tells that a result of XTree model must be a matching of nearest neighbour of keyword nodes.

Lemma 2. Let $c=LCA(a,b)$, and a is the nearest neighbour of b, or b is the nearest neighbour of a, then c must be a result of XTree model.

Proof. Assume after the excluding process, c does not contain a or b, then

(1) A node c' contains a and b, and c' is the sub node of c;
(2) A node c' contains a, and another node c'' contains b, both c' and c'' is the sub node of c.

In the first case, c' is the sub node of c and contains both a and b, then c can not be the LCA of a and b, it contradicts the definition of LCA.

In the second case, let $c'=LCA(a,b')$, $c''=LCA(a',b)$, then b' is closer to a and b, a' is closer to b than a. So neither b is nearest neighbour of a, nor a is the nearest neighbour of b, deriving a contradiction.

The discussion tells that c contains at least one keyword node after the excluding process. On the other side, $c=LCA(a,b)$, has contained both keywords, so c is a result of XTree model.

Lemma 2 tells that a matching of a nearest neighbor of keyword nodes must be a result of XTree model.

By Lemma 1 and Lemma 2, we get the following theorem.

Theorem 1. $Result(XTree)=MN$, where MN indicates the matchings of nearest neighbours of Keyword nodes.

Proof. By Lemma 1, we get $Result(XTree) \subseteq MN$. By Lemma 2, we get $MN \subseteq Result(XTree)$, so $Result(XTree)=MN$.

Theorem 1 tells that if we get all the matches of nearest neighbours of keyword nodes, we exactly get all the results of XTree model.

It is different from SLCA model that the XTree model does not follow associative law, but the case containing more keywords is similar. When the query contains only two keywords, a keyword node a should find its nearest neighbour b, while when the query contains m keywords, a keyword node kn_1 should find its $m-1$ nearest neighbours.

Before we present the query algorithm of XTree model, we should clarify the two definitions of "distance" in XML document trees. The precise definition of the "distance" of two nodes in XML document trees should be on the base of Dewey ID: The two nodes are closer if they have a longer common prefix of Dewey ID. However, the Dewey ID definition is too expensive for the implementations. The implementations usually choose another way, measuring the distance by the global order of nodes: The two nodes are closer if they are closer in the global order. The latter definition treats Dewey ID as a liner ID, and does not pay attention to the layered information contained in Dewey ID.

The two definitions get difference in the document trees shown in Figure 4. The Dewey ID definition thinks X_1 and Y_2 are closer, while the global order definition thinks X_1 and Y_1 are closer. However, the difference will not affect the correctness of our algorithm, we have the following theorem:

Theorem 2. Assume there are only two keywords, and a is a keyword node, b and b' are the other two keyword nodes which are the nearest neighbours of a, where b is under the definition of global order and b' is under the definition of Dewey ID. Let $c=LCA(a,b)$, $c'=LCA(a,b')$, then $c=c'$.

Proof. Assume $b \neq b'$, $DeweyID(c')=c'_1c'_2...c'_k$, $DeweyID(c)=c_1c_2...c_l$. Since b' is the nearest neighbor of a under the definition of Dewey ID, they should have a longer common prefix of Dewey ID, that is $l \leq k$.

Assume $l<k$, let $DeweyID(a)=c'_1c'_2...c'_k a_{k+1} a_{k+2}...$, $DeweyID(b)=c_1c_2...c_l b_{l+1} b_{l+2}...$, Because b is the nearest neighbor of a under the definition of global order, and $l<k$, then for b_{l+1}, there exists $|b_{l+1}-a_{l+1}| \leq |b'_{l+1}-a_{l+1}|$. On the other hand, we can get $b'_{l+1}=a_{l+1}=c_{l+1}$ by $DeweyID(c)$, that is $|b'_{l+1}-a_{l+1}|=0$, so $b_{l+1}=a_{l+1}$. Similarly, for b_{l+2}, b_{l+3} and b_k we get the same conclusion. So there must exist $l=k$ and $b_k=a_k$, that is $c=c'$.

Theorem 2 tells that although the global order definition may find the wrong match, it eventually get the right LCA result, so using the global order to measure the distance of keyword nodes will not effect the correctness of implementations.

By the discussion above, we have the algorithm, which can get all XTree nodes. We call it XScan, and Figure 2 shows the pseudo code.

XScan use the following steps to get all XTree nodes:

(1) Assume the query contains m keywords. First, XScan should get the global order of all keyword nodes mentioned in the query. It can be done effectively by the help of index.

(2) Initially, XScan maintains m points in *pArray* which points to the keyword nodes in m sets (line *3*).

(3) At the beginning, the points are all set to be the first elements in the set. Then XScan will choose the smallest keyword node as a target node (line *4-8*), and find the $m-1$ nearest neighbors of the target node. All the keyword nodes pointed by *pArray* are just bigger than the target node, so by the global order definition of distance, the nearest neighbor is the node *pArray* points, or the node before the node *pArray* points in the same set, that is, the nearest neighbor is *Keywords[i][pArray[i]]* or *Keywords[i][pArray[i]-1]* (line *9-19*).

(4) After getting the $m-1$ nearest neighbours, the LCA of the target node and the $m-1$ neighbours is a result of XTree model (line *20*).

(5) At the end of each loop, *pArray[i]* will step forward to prepare for the next loop (line *21*).

The key point of XScan is *pArray*. At the beginning of each loop, XScan chooses the smallest node from *pArray* and at the end of each loop, the corresponding point in *pArray* steps forward, this process makes sure that XScan always check the node by the global order. On the other hand, for each set *Keywords[i]*, there exists $*(pArray[i]-1) \leq target \leq *(pArray[i])$, then by the global order definition of the distance, the nearest neighbor of the target node is the node *pArray[i]* points, or the node before the node *pArray[i]* points, that is, the nearest neighbour will be chosen from only two nodes. So for every target node, the cost of searching the nearest neighbour is constant. So for a given query, the cost of XScan is $O(\sum n_i)$, where the query contains m keywords, and the size of the ith keyword set is n_i. XScan has a linear complexity.

```
1. function XScan( set[] Keywords, int size )
2. {
3.    int pArray = new int[size];
4.    for( int i = 0; i < size; ++i )
5.       pArray[i] = 0;
6.    while( true ){
7.       int minIndex = the index of the smallest node;
8.       Elem target =*pArray[minIndex];
9.       Elem result = target;
10.      for( int i = 0; i < size; ++i ){
11.         if( i == minIndex )
12.            continue;
13.         right = Keywords[i][pArray[i]];
14.         left = Keywords[i][pArray[i]-1];
15.         if( |left-base| < |right-base| )
16.            result = LCA( left, result );
17.         else
18.            result = LCA( right, result );
19.      }
20.      add result to result set;
21.      ++pArray[minIndex];
22.      if( all elemets in Keywords has been scaned )
23.         break;
24.   }
25.}
```

Fig. 2. The pseudo code of XScan

2.2   GenerateXTree: An Effective Algorithm for Generating Results

SLCA model and XRANK model return only nodes as results. Every result is thought as the root of a result tree. However, the result tree can be extremely large and it is not an easy job for users to find the corresponding keyword nodes.

Our XTree model will return comprehensive trees as results, which contain only XTree nodes and keyword nodes. After getting all XTree nodes, XTree model will connect them with corresponding keyword nodes to get result trees. This is different from traditional tree-returned algorithm such as MCT, which directly retrieve result trees from corpus.

The key problem here is to connect all nodes correctly at a linear cost. We have known that XScan check the keyword nodes in the global order, so we believe that there must be some ordered relationships between the generated XTree nodes.

Theorem 3. Let a, a' and b, b' be the keyword nodes in keyword set A and B, and $a<a'$, $b<b'$. Let $c=LCA(a,b)$, $c'=LCA(a',b')$, if the length of the Dewey IDs of c and c' are the same, then $c \leq c'$.

Proof. Let $DeweyID(c)=c_1c_2...c_m$, $DeweyID(c')=c'_1c'_2...c'_m$, for any i, c_i is the ith number in the Dewey IDs of c, a and b, and c'_i is the ith number in the Dewey IDs of c', a', b'. Since $a<a'$, $b<b'$, then $c_i \leq c'_i$. On the other hand, the length of Dewey IDs of c and c' are the same, so $c \leq c'$.

Theorem 3 tells that the Xtree nodes generated by XScan are layered ordered. The extra cost introduced is only that XScan needs to check whether the new generated node is equal to the previous one in the same layer.

If we put the keyword nodes in the corresponding level at the same time, we make all keyword nodes and XTree nodes be layered ordered. If we navigate all ordered nodes from top to bottom, we get the breadth-first traverse sequence of the result trees. It is a common problem to restore the original trees by their breadth-first traverse sequence. The pseudo code is shown in Figure 3.

The procedure, which we call GenerateXTree, has the following steps to restore the original result tree:

(1) Initially, GenerateXTree maintains two sets of nodes (line *3*). Set *current* stores the nodes generated in the current layer, and set *last* stores the nodes have been generated in bottom layers. Note that the nodes in set *last* may have already been connected to their sub nodes.

(2) GenerateXTree scans the sequence from bottom to top (line *4*). It compares the nodes in current layer and the nodes in *last* one by one (line *6*).

(3) If the node in *last* is smaller than the node in current layer, then it can not be the sub of after nodes, so GenerateXTree will add it to *current* (line *7-10*).

(4) If the node in *last* is bigger than the node in current layer, it means that the former may be the sub of the latter, if so, GenerateXTree will connect them together (line *13*). Since the node in current layer may be still the parent of the next node in *last*, the procedure will jump to the next loop.

(5) If the node in *last* is bigger than the node in current layer but the former is not the sub of the latter, it means that the node in current layer has found all its subs, so GenerateTree will add it to *current* (line *14-15*).

(6) At last, *current* will copy all its nodes to *last* to prepare for the next loop.

To restore the original result tree, GenerateXTree needs to scan all keyword nodes and XTree nodes once. It also has a linear complexity. Usually, if the nearest neighbors are not in the same document, we will exclude them from the result set. So the size of result set is usually much less than the total size of original keyword sets, which means that GenerateXTree is usually much faster than XScan.

The pseudo code of GenerateXTree tells that there are only XTree nodes and keyword nodes in the result tree. This can help users easily find the needed information. But the tree can still be too large to navigate. In the implementation, we set a threshold k, if the number of nodes in a result tree is bigger than k, we will break it into several small trees. Each small tree uses a sub XTree node as the root, and the original tree will degenerate to an index that uses all the sub roots as its leaf nodes.

```
1.  function GenerateXTree( set[] nodes, int depth )
2.  {
3.     set current, last;
4.     for( int i = depth-1; i >=0; --i ){
5.        int p1 = p2 = 0;
6.        while( p1<last.size() && p2<nodes[i].size() ){
7.           if( last[p1] < nodes[i][p2] ){
8.              current.add( last[p1] );
9.              ++p1;
10.          }
11.          else{
12.             if( last[p1] is the sub of nodes[i][p2] )
13.                nodes[i][p2].addsub( last[p1] );
14.             else
15.                current.add( nodes[i][p2] );
16.             ++p2;
17.          }
18.       }
19.       if( p1==last.size() && p2<nodes[i].size() ){
20.          for( int j = p2; j < nodes[i].size(); ++j )
21.             cuurent.add( nodes[i][j] );
22.       else if( p1<last.size() && p2==nodes[i].size() )
23.          similar with the previous case;
24.       last = current;
25.    }
26.    return last;
27.}
```

Fig. 3. The pseudo code of GenerateXTree

3 Experimental Evaluation

We implement an XML keyword retrieval system in C++ using the Apache Xerces XML parser [14] and Berkeley DB [15]. We use DBLP as the data set, which contains 12,881,441 nodes and 388,826 keywords. The max depth of DBLP is 6 and the most frequent keyword appears in 1,810,451 nodes. We evaluate the performance of XScan and GenerateXTree on the different number of keywords and different size of keyword sets. The platform is a desktop with 1.66G*2 CPU and 1G RAM. The results are shown in following Figures.

Figure 4 shows the performance of XScan and GenerateXTree when the query contains only two keywords. The part (a) is the performance when the sizes of both keyword sets are the same. The time cost by XScan is 31, 265 and 3719 milliseconds respectively when the size of each keyword set is 10K, 100K and 1M. The part (b) of Figure 4 is the performance when we fix the size of one keyword set as 1M and vary the size of another keyword set. The time cost by XScan is 1281, 1594 and 3719 milliseconds respectively when the size of another keyword set is 10K, 100K and 1M. We can see that the performance of XScan is determined by the total size of all

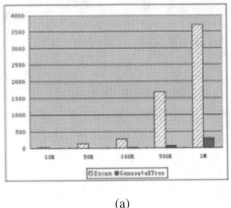

(a)

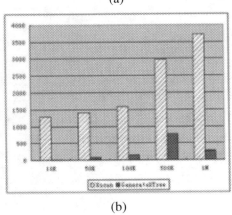

(b)

Fig. 4. The performance of XScan and GenereateXTree when the number of keywords is 2

keywords. That is, the most frequent keywords in the query are the determinant factors. The time cost by GenerateTree shown in Figure 4 is much less than XScan as we expected. It is determined by the size of result set, and in this example, it decreases while the size of another keyword set increase from 500K to 1M.

Figure 5 shows the performance of XScan and GenerateTree when the query contains three keywords. The part (a) is the performance when the sizes of all keyword sets are the same. The time cost by XSCan is 32, 406 and 9031 milliseconds when the size of each keyword set is 10K, 100K and 1M respectively. The part (b) is the experimental result in the case that we fix the size of one keyword set as 1M and vary the size of the other two keyword sets. The time cost by XScan is 1344, 1985 and 9031 milliseconds when the sizes of the other two keyword sets are 10K, 100K and 1M respectively. The part (c) is the experimental result in the case that we fix the size of two keyword sets as 1M and vary the size of the left keyword set. The time cost by XScan is 5625, 6093 and 9031 milliseconds when the size of the left keyword set is 10K, 100K and 1M respectively. Figure 5 tells the same thing as Figure 4, the performance of XScan is determined by the total size of keyword sets and the most frequent keywords in the query are determinant factors. Figure 5 also tells that XScan also has a good scalability on the total size of keyword sets when the number of key-words increases to three. Both Figure 4 and 5 tell that XScan has a good scalability on the total size of all keyword sets no matter how many keywords the query contains. This is because that the main cost of XScan is to get the LCA of two keyword nodes. XScan will repeat this operation for millions times though the cost of single one is tiny.

We also find that XScan has a good scalability on the number of keywords, as shown in Figure 6. In the experiment, the sizes of keyword sets are all set as 50K. XScan costs 125, 203, 265, 328 and 406 milliseconds respectively when the query contains 2, 3, 4, 5 and 6 keywords. This reason is that the performance of XScan is determined by the number of LCA operations, and m keywords need $m-1$ groups of LCA operations. So the relationship between the performance of XScan and the number of keywords is linear.

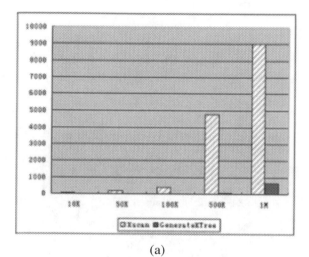

(a)

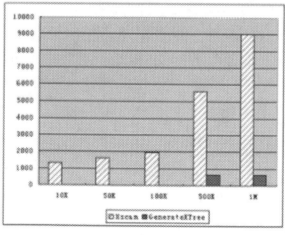

(b)

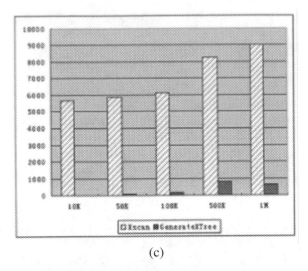

(c)

Fig. 5. The performance of XScan and GenereateXTree when the number of keywords is 3

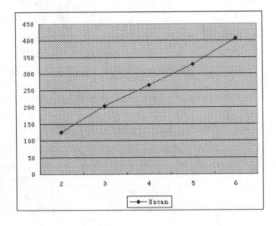

Fig. 6. The performance of XScan with different numbers of keywords

4 Conclusions

In this paper, we present a new XML keyword retrieval model, XTree, which can cover every keyword node and return the comprehensive result trees. For this model, we also propose an efficient query algorithm, Xscan, and an efficient result construction algorithm, GenerateTree. The extensive experiments show that XScan and GenerateTree have good scalabilities on both the number of keywords and the number of keyword nodes. However, XTree model is still slower than traditional retrieval models such as SLCA model and XRANK model because it returns much more nodes.

For future work, there is a lot of interesting research issues. First, the main cost of XTree is the LCA operation. This is because that the Dewey ID is compressed deposited and it should be decoded and encoded in the process. In the future, we will try to find more efficient algorithms, which can directly manipulate the compressed format of Dewey ID, to enhance the whole performance of XTree. Second, the structural information contained in XML document trees is so important that SLCA model only returns the most specific nodes. Our XTree model tries to cover every keyword node and return every possible result. So, we need an effective ranking method that can help users distinguish different nodes and find their needed information quickly in the massive number of results.

References

1. Clark, J., DeRose, S.: XML Path Language (XPath) 1.0 (November 1999), http://www.w3.org/TR/xpath
2. XQuery 1.0: An XML Query Language (June 2001), http://www.w3.org/XML/Query
3. Shanmugasundaram, J., Tufte, K., Zhang, C., Gang, H., DeWitt, D.J., Naughton, J.F.: Relational databases for querying XML documents: Limitations and opportunities. In: VLDB (1999)
4. Kanne, C.C., Moerkotte, G.: Efficient Storage of XML Data. In: ICDE (2000)
5. Schmidt, A., Kersten, M.L., Windhouwer, M., Waas, F.: Efficient relational storage and retrieval of XML documents. In: Suciu, D., Vossen, G. (eds.) WebDB 2000. LNCS, vol. 1997, p. 137. Springer, Heidelberg (2001)
6. Cooper, B.F., Sample, N., Franklin, M.J., Hjaltason, G.R., Shadmon, M.: Proc. A Fast Index for Semistructured Data. In: VLDB (2001)
7. Chien, S.Y., Vagena, Z., Zhang, D.H., Tsotras, V.J., Zaniolo, C.: Efficient Structural Joins on Indexed XML Documents. In: VLDB (2002)
8. McHugh, J., Widom, J., Abiteboul, S., Luo, Q., Rajaraman, A.: Indexing Semistructured Data. Technical Report (1998)
9. Bohannon, P., Freire, J., Roy, P., Simeon, J.: From XML schema to relations: A cost-based approach to XML storage. In: ICDE (2002)
10. Cho, J., Rajagopalan, S.: A fast regular expression indexing engine. In: ICDE (2002)
11. Guo, L., Shao, F., Botev, C., Shanmugasundaram, J.: XRANK: Ranked Keyword Search over XML Documents. In: Proceedings of SIGMOD, pp. 16–27 (2003)
12. Xu, Y., Papakonstantinou, Y.: Efficient Keyword Search for Smallest LCAs in XML Databases. In: Proceedings of SIGMOD, pp. 527–538 (2005)
13. Liu, Z., Chen, Y.: Identifying Meaningful Return Information for XML Keyword Search. In: Proceedings of SIGMOD, pp. 329–340 (2007)
14. Xerces-C, http://xerces.apache.org/xerces-c/
15. Berkeley DB, http://www.oracle.com/technology/products/berkeley-db/index.html

Effective Top-k Keyword Search in Relational Databases Considering Query Semantics

Yanwei Xu[1,2], Yoshiharu Ishikawa[1], and Jihong Guan[2]

[1] Graduate School of Information Science, Nagoya University, Japan
[2] School of Electronics and Information Engineering, Tongji University, China

Abstract. Keyword search in relational databases has recently emerged as a new research topic. As a search result is often assembled from multiple relational tables, existing IR-style ranking strategies can not be applied directly. In this paper, we propose a novel IR ranking strategy considering query semantics for effective keyword search. The experimental results on a large-scale real database demonstrate that our method results in significant improvement in terms of retrieval effectiveness as compared to previous ranking strategies.

Keywords: top-k, keyword search, effective, relational database.

1 Introduction

With the amount of available text data in relational databases growing rapidly, the need for ordinary users to effectively search such information is increasing dramatically. Keyword search is the most popular information retrieval method because the user needs to know neither a query language nor the underlying structure of the data. Keyword search in relational databases has recently emerged as an active research topic. In this paper, we focus on how to support effective top-k keyword search in relational databases.

Although most of the popular DBMSs support full-text search, they only provide support for retrieving tuples relevant to a query within the same relation. A unique feature of keyword search in relational databases is that search results are often joined tuples from multiple relations.

Example 1: Suppose a user wants to search papers written by *"Ralf Steinmetz"* with *"p2p"* in their titles from the DBLP[1] database (its schema is shown in Figure 1). He might give a query containing two keywords: *"p2p Steinmetz"*. Our system will return the results shown in Table 1, where relevant tuples from multiple relations (presented in bold font) are joined together to form a meaningful answer to the query. Table 1 shows that three papers with *"p2p"* in their titles were written by *"Ralf Steinmetz"*.

Recently, there have been many studies dedicated to keyword search in relational databases [1,2,3,4,5]. Among these, [3] was the first to consider top-k keyword search in relational databases; it incorporates a state-of-the-art IR ranking

---

[1] http://dblp.mpi-inf.mpg.de/dblp-mirror/index.php

L. Chen et al. (Eds.): APWeb and WAIM 2009, LNCS 5731, pp. 172–184, 2009.

Table 1. Top-3 results for query *"p2p Steinmetz"*

1 **Article:**Token-Based Accounting for P2P-Systems.→**Author:** Ralf Steinmetz
2 **Article:**An Adaptable, Role-Based Simulator for P2P Networks.→ **Author:**Ralf Steinmetz
3 **Article:**Self-protection in P2P Networks: Choosing the Right Neighbourhood.→ **Author:**Ralf Steinmetz

formula to address the retrieval effectiveness issue and presents several efficient query execution algorithms optimized for returning top-k relevant answers. [4] improves the ranking formula in [3] by adapting four normalizations. [5,6] further modify the ranking formula of [3] by introducing the concept of a *virtual document* and present two efficient query evaluation algorithms for their ranking formula. [7] takes another approach to address keyword search based on the Steiner tree.

Due to the fuzzy nature of keyword queries, result ranking is vital for retrieval effectiveness. Despite the results from previous studies, there are still several issues with existing ranking methods, some of which may discourage users to use keyword search systems. In this paper, we present a method for improving the ranking formula by considering query semantics.

The main contributions of this paper are as follows:

- We introduce the concept of query semantics for keyword search in relational databases. Although this concept has been mentioned in previous works [5], it was not considered as a factor for ranking search results.
- We propose a method for incorporating query semantics into the ranking formulas proposed in [3,5]. To our knowledge, our paper is the first to rank CNs.[2]
- We conduct comprehensive experiments on large-scale real databases. The experimental results show that our approach is better than existing ones in terms of effectiveness.

The rest of the paper is organized as follows: Section 2 presents the basic concepts and the method for generating the relevant answers of a query. Section 3 introduces the ranking strategies used in previous works. Section 4 presents the concept of query semantics and our method of ranking answers by considering query semantics. Section 5 shows the experimental results. Section 6 discusses the related works. Section 7 concludes this paper.

2   Preliminaries

In this section, we describe the framework for generating answers for a given keyword query. Section 2.1 describes some basic concepts such as Candidate Network (CN) and Joint-Tuples-Tree (JTT). We follow the definitions of previous work [5,8]. Section 2.2 describes the framework of generating query answers.

---

[2] CN is short for *candidate network*, which will be introduced in Section 2.

2.1   Basic Concepts

We first define some terms used throughout the paper. A relational database composed of a set of relations $R_1, R_2, \cdots, R_n$. A *Schema Graph* (*SG*) is a directed graph with the relations as its nodes and the foreign key to primary key relationships of the relations as its edges. Figure 1 shows the schema graph of DBLP used in this paper. A *Joint-Tuple-Tree* (*JTT*) T is a joining tree of different tuples. Each node t_i is a tuple in the database, and each pair of adjacent tuples in T is connected via a foreign key to primary key relationship. The three results in Table 1 are examples of JTTs. A JTT is an answer to a keyword query if it contains more than one keyword of the query and each of its leaf tuples must contain at least one keyword. A *Query Tuple Set* R^Q is a set of all tuples which belong to relation R; these tuples contain at least one keyword of the query Q. We call R^F the *free tuple set*, which is the set of all tuples in relation R and we use R^* to denote a *tuple set*, which can be either a query tuple set or a free tuple set. A *Candidate Network* (*CN*) is a tree of tuple sets R^Q or R^F with the restriction that every leaf node must be a query tuple set. Every edge (R_i^*, R_j^*) in a CN corresponds to an edge (R_i, R_j) in the schema graph *SG*. A CN can be easily transformed into its equivalent SQL statement and executed through the DBMS. The *size* of a CN is the number of its tuple sets.

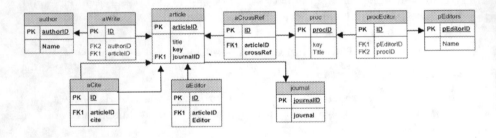

Fig. 1. DBLP schema graph

2.2   Answer Generation

Given a keyword query, the system first generates all the non-empty query tuple sets R^Q for all the relations R. These non-empty query tuple sets and the schema graph are inputted to the CN generator to generate all the valid CNs. For this purpose, [8] has proposed a breadth-first algorithm that is both sound and complete. It can enumerate all the CNs of size no more than a specified number without violating any pruning rules. There are three pruning rules used in [5], which are listed below. We show the traces of the CN generation algorithm for query *"p2p Steinmetz"* in Example 1 in Table 2 (for simplicity, suppose there are only two non-empty query tuple sets $article^Q$ and $author^Q$, and omit relation *aCite*).

Rule 1. Prune duplicate CNs
Rule 2. Prune non-minimal CNs i.e., CNs with free tuple sets as leaf nodes

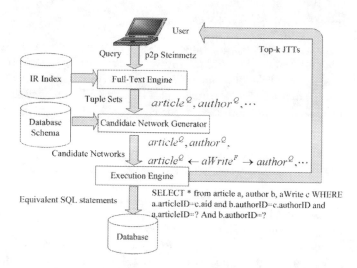

Fig. 2. Query processing framework

Table 2. Enumerating CNs for query *"p2p Steinmetz"*

Size	CN ID	CN	Valid?
1	CN_1	$article^Q$	Y
1	CN_2	$author^Q$	Y
2		$article^Q \leftarrow aWrite^F$	n
2		$article^Q \rightarrow journal^F$	n
2		$article^Q \leftarrow aCrossRef^F$	n
3	CN_3	$article^Q \leftarrow aWrite^F \rightarrow author^Q$	Y
3		$article^Q \leftarrow aWrite^F \rightarrow author^F$	n
3	CN_4	$article^Q \rightarrow journal^F \leftarrow article^Q$	Y
3		$article^Q \rightarrow journal^F \leftarrow article^F$	n
3		$article^Q \leftarrow aCrossRef^F \rightarrow proc^F$	n
3		$author^Q \leftarrow aWrite^F \rightarrow article^Q$	n
4	$\vdots$ $\vdots$		$\vdots$

Rule 3. Prune CNs of type: $R^Q \leftarrow S^* \rightarrow R^Q$. The rationale is that every resulting JTT would contain the same tuple from R^Q for two times.

Finally, the generated CNs are evaluated to identify the top-k query results based on some relevance formulas. Figure 2 shows query processing framework, which is a modified version of [3].

3   Ranking Strategy

In this section, we first present the ranking strategies of previous works, then motivate our work by presenting an observation that reveals a problem in existing schemes.

3.1   Existing Ranking Strategies

Due to the fuzzy nature of keyword queries, result ranking is vital for retrieval effectiveness. The initial attempt was to simply rank results according to the size of JTTs [2,8]. Later, IR-Style [3] proposed a ranking formula based on a state-of-the-art IR scoring function e.g., formulas based on the TF-IDF weighting. The basic idea of the ranking method used in [3] is:

1. Assign to each tuple in the JTT a score by using a standard IR-ranking formula[3]; and
2. Combine the individual scores together by using a monotonic aggregation function to obtain the final score.

[4] suggested four sophisticated normalizations to the scoring function in [3]: tuple tree size normalization, document length normalization, document frequency normalization and inter-document weight normalization. The scoring function of [4] is not monotonic due to the four normalizations, and therefore the optimized query evaluation algorithms in [3] cannot be applied.

SPARK [5] models the entire JTT as a virtual document while the entire results produced by a CN is modeled as a document collection. SPARK computes the relevance score for a JTT T as follows:

$$score(T,Q) = score_a(T,Q) \cdot score_b(T,Q) \cdot score_c(T,Q), \tag{1}$$

$$score_a(T,Q) = \sum_{w \in T \cap Q} \frac{1 + ln(1 + ln(tf_w(t)))}{1 - s + s \cdot \frac{dl_T}{avdl(CN^*(T))}} \cdot ln(idf_w), \tag{2}$$

where $tf_w(T) = \sum_{t \in T} tf_w(t)$, $idf_w = \frac{N(CN^*(T))+1}{df_w(CN^*(T))}$,

$$score_b(T,Q) = 1 - \left(\frac{\sum_{1 \leq i \leq m}(1 - T.i)^p}{m}\right)^{\frac{1}{p}}, \tag{3}$$

where $T.i = \frac{tf_{w_i}(T)}{max_{1 \leq j \leq m} tf_{w_j}(T)} \cdot \frac{idf_{w_i}}{max_{1 \leq j \leq m} idf_{w_j}}$,

$$score_c = (1 + s_1 - s_1 \cdot size(CN)) \cdot (1 + s_2 - s_2 \cdot size(CN^{nf})), \tag{4}$$

where $tf_w(t)$ denotes the number of instances of w in t, dl_T denotes the length of all the text attributes of T, $CN(T)$ denotes the CN T belongs to, $CN^*(T)$ is identical to $CN(T)$ with the exception that each tuple set is free, $avdl(CN^*(T))$ is the average length of JTTs for $CN^*(T)$, $N(CN^*(T))$ denotes the number of JTTs for $CN^*(T)$, and $size(CN^{nf})$ is the number of non-free tuple sets for the CN. $score_a$ is an IR-style ranking score based on the TF-IDF weighting. $score_b$ acts as the completeness factor and gives biases toward the JTTs which contain

---

[3] This score is often automatically computed by the DBMS by using the full-text indexing engine.

all of the keywords in query Q to those which only contain a few keywords. The tuning parameter p in Eq.(3) can smoothly switch the completeness factor biased towards the OR semantics to the AND semantics. $score_c$ is a JTT size factor and its degree of penalties for large CN is between [3] and [4]. The ranking function of SPARK addresses an important deficiency in existing methods and results in substantial improvement of the quality of search results [5].

3.2   Problems with Existing Ranking Functions

The vector space model based on TF-IDF weighting is used to compute the relevance between a keyword query and documents from a document collection. In the setting of the keyword search in relational databases, there will be multiple document collections (each collection is either a relation or a CN). The above scoring functions only compute a document's relative relevance to the query in the document collection it belongs to. However, these document collections have different levels of importance to the query, and therefore the final score of a document must reflect the importance of the document collection it belongs to. For example, the top ten results of the query in Example 1 returned by SPARK with $p = 2$ (a value of 2.0 is already good enough to enforce the AND-semantics [5]) are listed in Table 3. We can see that most of the results are not useful to the user: no papers written by *"Ralf Steinmetz"* with *"p2p"* in the title are returned.

Table 3. Top ten results for query *"Steinmetz p2p"* by SPARK

JTT	Score	CN
E. Steinmetz	3.40	$author^Q$
Uli Steinmetz	3.37	$author^Q$
2 P2P or Not 2 P2P?	3.35	$article^Q$
2 P2P or Not 2 P2P?	3.35	$article^Q$
Ralf Steinmetz	3.34	$author^Q$
Arnd Steinmetz	3.34	$author^Q$
Rita Steinmetz	3.34	$author^Q$
Aase Steinmetz	3.34	$author^Q$
Oliver Steinmetz	3.28	$author^Q$
Ulrich Steinmetz	3.28	$author^Q$

The total number of tuples in every relation that contain the two keywords in DBLP are shown in Table 4, and enumerated CNs whose size is less than 4 are listed in Table 5.

Table 4. Statistics of keyword *Steinmetz* and *p2p*

Relation	Column	Keyword	Count
proc	title	P2P	11
article	title	P2P	1855
procEditor	Name	Steinmetz	1
author	author	Steinmetz	20

Table 5. Enumerated CNs for query *"Steinmetz p2p"*

CN ID	CN
CN_1	$article^Q$
CN_2	$author^Q$
CN_3	$procEditor^Q$
CN_4	$proc^Q$
CN_5	$article^Q \leftarrow aCrossRef^F \rightarrow proc^Q$
CN_6	$article^Q \leftarrow aWrite^F \rightarrow author^Q$
CN_7	$article^Q \rightarrow journal^F \leftarrow article^Q$
CN_8	$pEditors^Q \leftarrow procEditor^F \rightarrow proc^Q$

Table 4 shows that the two keywords *Steinmetz* and *p2p* mostly occur at relation *author* and *article*, respectively. From a human perspective, results for CN_6 should be ranked higher than results from other CNs on the basis of the data in Table 4, even if we do not know the user's intentions. Unfortunately, *p2p* is such a popular keyword in *article* as compared to *Steinmetz* in *author* that idf_{p2p} in Eqs. (2) and (3) is very small as compared to $idf_{Steinmetz}$. As a result, answers from CN_1 and CN_2 containing *Steinmetz* are ranked as the top ten answers as Table 3 shows. Of course, $score_a$ and $score_b$ of the answers for CN_6 will be larger than the answers for CN_1 and CN_2. However, the degree of increase is very small because $idf_{p2p} \ll idf_{Steinmetz}$ and is counteracted by the decrease of $score_c$, as the JTT size is 3.

We will show our solution to this problem in next section.

4   Ranking with Query Semantics

In this section, we will first introduce the concept of query semantics, and then discuss how to use it to improve the ranking strategy. Our proposed method shows a notable improvement in the effectiveness of keyword search.

4.1   Query Semantics

We believe that the problem in Table 3 is caused by the fact that the keyword search system does not understand the user's true intention. This is why some keyword search systems allow a query to contain database schema data such as "author:Steinmetz". However, requiring an ordinary user to write queries containing database schema data is not realistic and violates the original motivation for keyword search systems.

In actual commercial databases, there is always a large number of relations. Due to the E-R model and the normalization requirement, each relation stores information of a certain kind of entities, and hence it has its own special keyword set. For example, keywords in relation *author* in DBLP are unlikely to occur in relation *article* or *proc*.

When a user inputs a short query, we can assume that there is a strong possibility that he has a preference for the relation selection for each keyword. For example, he prefers the relation *author* to relation *article* when he inputs the name of person. The implicit relationship between keywords and relations specifies the hidden user's intention for the query.

We refer to such relation preference of keywords as the *semantics* of the query. If a CN contains all the relations in the semantic set of a query, results from this CN will be more relevant to the query.

Example 2: a query contains an author name *Steinmetz* and a research area keyword *p2p*, which shows that the user wants to search papers about *p2p* that were written by an author named *Steinmetz*. Hence, the semantic set of query "*Steinmetz p2p*" is $\{article, author\}$. JTTs for CN_6 in Table 5 are more relevant to the query "*Steinmetz p2p*" than JTTs for CN_8.

If the semantics for a keyword query can be obtained exactly and used to rank query results, query effectiveness can be drastically improved. For example, the papers with *p2p* in the title written by *Ralf Steinmetz* will be ranked at the top of the answers. We present our method for obtaining the query semantics and using it to rank answers in the next section.

4.2   Incorporating Query Semantics into Ranking

We propose modeling a relation as a document, in which case tuples in relations will be modeled as words or sentences. Consequently, the database is a document collection composed of relations as documents. By adopting such a model, we can naturally compute the IR-style relevance score of each relation for a keyword.

For a keyword w, we can easily find all the relations $R_1, R_2, \cdots, R_t$ that have tuples containing w and the corresponding numbers of tuples by using a full-text index. Then, for each $R_i (1 \leq i \leq t)$, we use the following formula, which is based on TF-IDF weighting, to compute its relevance to keyword w:

$$p_w(R_i) = \frac{p_0 + ln(1 + ln(1 + df_w(R_i)))}{(1 - s) + s \cdot ln(1 + \frac{tc_{R_i}}{avtc})}, \tag{5}$$

where p_0 indicates the initial preference score for a keyword to an arbitrary relation, $df_w(R_i)$ is the number of tuples of R_i that contains keyword w, tc_{R_i} is the number of total tuples in R_i, $avtc$ is the average number of tuples of the relations in the database.

p_0 and s in Eq.(5) acts as two tuning parameters: small p_0 give higher preference to relations that have a large $df_w(R_i)$, while larger s gives higher preference to relations that have a small number of tuples. In our experiments, p_0 is tuned between 0.2 and 1 and s is tuned between 0.1 and 0.5. We observed that $p_0 = 0.6$ and $s = 0.2$ is appropriate for all the tested queries.

Then the preference of CN for a query Q is computed as:

$$score_s(CN, Q) = \sum_{w \in Q} max_{R_i \in CN} p_w(R_i). \tag{6}$$

We refer to $score_s(CN, Q)$ as the *semantic score* of a CN.

Finally, the relevance score of a JTT T to a keyword query Q is computed as:

$$SCORE(T,Q) = score(T,Q) \cdot score_s(CN(T),Q), \qquad (7)$$

where $score(T,Q)$ is computed by Eq.(1).

The calculated $score_s$ for the eight CNs in Table 5 are listed in Table 6 with $p_0 = 1.0, s = 0.25$. We also suggest the application of a modification to Eq.(3) as

$$T.i = \frac{ln(1 + tf_{w_i}(T))}{ln(1 + max_{1 \leq j \leq m} tf_{w_j}(T))} \cdot \frac{ln(idf_{w_i})}{ln(max_{1 \leq j \leq m} idf_{w_j})}, \qquad (8)$$

to increase the impact of a keyword w that has a small idf (e.g., keyword $p2p$). Table 7 shows the top ten results of SPARK for query "$p2p$ Steinmetz" with $p = 1.8$ and our modification when computing $score_b$. We find that the six results which belong to CN_6 are ranked at the top of the results. More importantly, the result belonging to CN_8 is ranked appropriately.

Table 6. Calculated $score_s$ of the eight CNs

CN_1	CN_2	CN_3	CN_4	CN_5	CN_6	CN_7	CN_8
3.24	2.59	3.04	2.14	3.44	4.66	3.24	3.7

Table 7. Top ten results for query "$p2p$ Steinmetz"

JTT	CN ID	score
Token-Based Accounting for P2P-Systems.→ Ralf Steinmetz	CN_6	33.25
An Adaptable, Role-Based Simulator for P2P Networks.→ Ralf Steinmetz	CN_6	32.28
Self-protection in P2P Networks: Choosing the Right Neighbourhood.→ Ralf Steinmetz	CN_6	31.29
Globase.KOM - A P2P Overlay for Fully Retrievable Location-based Search.→ Ralf Steinmetz	CN_6	30.88
Ralf Steinmetz→ Proceedings P2P'08, Eighth International Conference on Peer-to-Peer Computing, 8-11 September 2008, Aachen, Germany	CN_8	30.78
Overlay Design Mechanisms for Heterogeneous Large-Scale Dynamic P2P Systems.→Ralf Steinmetz	CN_6	30.62
Working Group Report on Managing and Integrating Data in P2P Databases.→ Rita Steinmetz	CN_6	30.55
E. Steinmetz	CN_2	16.03
Uli Steinmetz	CN_2	15.89
Arnd Steinmetz	CN_2	15.75

5   Experiments

In this section, we experimentally discuss the impact of our proposed method on the effectiveness of top-k keyword search. We incorporate the semantic score into the ranking strategies of [3,5], and then compare its impact to the effectiveness.

5.1   Experimental Settings

Database: For our evaluation,we use the DBLP data set, which we decomposed into relations from a downloaded XML file according to the schema shown in

Table 8. Statistics of DBLP database

Relation Schema	# Tuples
article(articleID,key,title,journalID,$\cdots$)	1,092,239
aCite(id,articleID,cite)	109,625
author(authorID,author)	658,461
aWrite(id,articleID,authorID)	2,752,673
journal(journalID,journal)	730
proc(procID,key,title,$\cdots$)	11,108
pEditors(pEditorID,Name)	12,001
procEditor(id,procEditorID,procID)	23,540

Figure 1. We use many relations in order to represent the original data in the XML file as closely as possible. The size of the XML file is 478MB. Table 8 shows the basic statistics after the decomposition.

Query Set: We manually picked a large number of queries for evaluation. We attempted to include a wide variety of keywords and their combinations in the query set, such as the selectivity of keywords, the size of the most relevant answers, the number of potential relevant answers, etc. We focus on 20 queries with query length ranging from 2 to 4.

5.2 Measures

To measure the effectiveness, we adopt two metrics used in previous studies [4,5]: *a)* number of top-1 answers that are relevant (#**Rel**), and *b)* *reciprocal rank* (**R-Rank**), for a given query. The reciprocal rank is 1 divided by the rank at which the first correct answer is returned or 0 if no correct answers are returned.

In order to identify the relevant answers for each query, we used all the ranking strategies ([3],[5] and ours) for each query and merged their top-50 results. Then, we manually evaluated the results and selected the relevant answer(s) for each query.

5.3 Results and Discussion on Effectiveness

We show the #**Rel** of [3,5] and our proposed method on the DBLP dataset in Table 9. Figure 3 and 4 show the *reciprocal ranks* of 13 and 10 queries, respectively. We use [3](S), [5](S) to denote that the relevance score is computed by considering the semantic score of CNs. M in [5](M) and [5](SM) denotes the modification shown in Eq.(8). $p = 1.5(2)$ denotes the parameter p in Eq.(3) is set to 1.5(2). From Table 9, we can see the notable improvement brought by our method to the ranking strategies of [3,5] in terms of effectiveness. [5](SM) can always return the relevant answer(s) for a query. We also find that the semantic score works well with the method of [3]. Although the #**Rel** of [3](S) is not large, we can find relevant answers in the Top-20 answers returned by [3](S) in most cases, as shown in Figure 4.

Table 9. Impacts on **#Rel**

$[5](p=2)$	$[5](M)(p=1.5)$	$[5](M)(p=2)$	$[5](S)(p=1.5)$	$[5](SM)(p=1.5)$	$[3]$	$[3](S)$
4	17	18	12	19	0	5

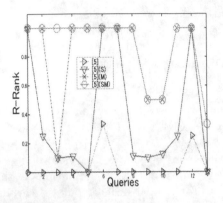

Fig. 3. Impacts on **R-Rank** (1)   **Fig. 4.** Impacts on **R-Rank** (2)

The experimental results in Table 9 and Figure 3 show that Eq.(8) is characterized by a notable improvement as compared to the method of [5] in terms of **#Rel** and *reciprocal ranks*. There are two reasons for this: Eq.(8) produces a stronger bias to answers that contain all of the keywords in a query; our manually evaluated relevant answers are based on the **AND** semantics for queries. Although they have similar **#Rel** and **R-Rank**, there is a great difference between $[5](S)$ and $[5](SM)$ in the structure of the top-k answers as compared to $[5](M)$: there is always a larger CN number. For example, $[5](M)$ ranks all the answers for CN_6 at the top for the query *"p2p Steinmetz"*. However, we can also find answers for CN_1, CN_2 and CN_8 in the top-ten answers returned by $[5](S)$ and $[5](SM)$. Therefore, although we believe users might target different results with their queries, a larger CN number can meet the demands of more users.

6   Related Work

Keyword search in relational databases has recently emerged as a new research topic [9]. Existing approaches can be broadly classified into two categories: those based on candidate networks [2,3,8] and others based on Steiner trees [1,10,11].

Mragyati [12], Discover [8], DBXplorer [2], IR-Style [3] and ObjectRank [13] are several early keyword search systems for relational databases. Discover and DBXplorer only rank tuple trees according to their sizes. IR-Style proposed ranking of tuple trees according to their IR relevance scores to a query. Our work adopt the same framework with [2,3,4,8], and can be viewed as a further improvement along the line of enhancing the retrieval effectiveness. ObjectRank and

ObjectRank2 [14] apply authority-based ranking to keyword search in databases modeled as labeled graphs. Authority originates at the nodes (objects) containing the keywords and flows to objects according to their semantic connections.

Banks [1] also finds tuple trees from the data graph directly by using the Steiner tree algorithm. For a data graph, it uses PageRank style methods to assign weights to tuples and edges between them. Banks2 [10] is an improvement of Banks which introduces a novel technique of bidirectional expansion to improve search efficiency. Li et al. [7] proposed a new concept referred to as a *compact Steiner Tree*, which can be used to approximate the Steiner tree problem for answering top-k keyword queries efficiently. They also proposed a novel structure-aware index to support keyword search.

Most recently, keyword search has been studied in a few generalized contexts as well [11,15]. [15] describes a solution to the keyword-search problem over heterogeneous relational databases. The scoring function of [15] is adapted from [3] by adding two more equally weighted terms that capture the match confidence of *FK joins* and the corresponding attribute value pairs in an answer, which may be a JTT composed of tuples come from multiple databases.

7   Conclusions

Keyword search allows non-expert users to find text information in relational databases with much higher flexibility. In this paper, we proposed a novel ranking strategy for effective keyword search considering query semantics. Our method can solve the problems with the ranking strategies proposed in previous works. We also presented a modification of an existing ranking strategy. Our method can be incorporate into existing ranking strategies and does not require excessive additional computation. The results of experiments performed on a large-scale real dataset show that our method results in a significant improvement in terms of retrieval effectiveness.

Acknowledgement

This research was partly supported by the Grant-in-Aid for Scientific Research, Japan (#19300027).

References

1. Aditya, B., Bhalotia, G., Chakrabarti, S., Hulgeri, A., Nakhe, C., Parag, S.S.: BANKS: Browsing and keyword searching in relational databases. In: VLDB, pp. 1083–1086 (2002)
2. Agrawal, S., Chaudhuri, S., Das, G.: DBXplorer: enabling keyword search over relational databases. In: ACM SIGMOD, p. 627 (2002)
3. Hristidis, V., Gravano, L., Papakonstantinou, Y.: Efficient ir-style keyword search over relational databases. In: VLDB, pp. 850–861 (2003)

4. Liu, F., Yu, C., Meng, W., Chowdhury, A.: Effective keyword search in relational databases. In: ACM SIGMOD, pp. 563–574 (2006)
5. Luo, Y., Lin, X., Wang, W., Zhou, X.: SPARK: top-k keyword query in relational databases. In: ACM SIGMOD, pp. 115–126 (2007)
6. Luo, Y., Wang, W., Lin, X.: SPARK: A keyword search engine on relational databases. In: ICDE, pp. 1552–1555 (2008)
7. Li, G., Feng, J.-H., Lin, F., Zhou, L.-z.: Progressive ranking for efficient keyword search over relational databases. In: Gray, A., Jeffery, K., Shao, J. (eds.) BNCOD 2008. LNCS, vol. 5071, pp. 193–197. Springer, Heidelberg (2008)
8. Hristidis, V., Papakonstantinou, Y.: DISCOVER: Keyword search in relational databases. In: VLDB, pp. 670–681 (2002)
9. Wang, S., Zhang, K.: Searching databases with keywords. J. Comput. Sci. Technol. 20(1), 55–62 (2005)
10. Kacholia, V., Pandit, S., Chakrabarti, S., Sudarshan, S., Desai, R., Karambelkar, H.: Bidirectional expansion for keyword search on graph databases. In: VLDB, pp. 505–516 (2005)
11. Li, G., Ooi, B.C., Feng, J., Wang, J., Zhou, L.: EASE: an effective 3-in-1 keyword search method for unstructured, semi-structured and structured data. In: ACM SIGMOD, pp. 903–914 (2008)
12. Sarda, N.L., Jain, A.: Mragyati: A system for keyword-based searching in databases. CoRR cs.DB/0110052 (2001)
13. Balmin, A., Hristidis, V., Papakonstantinou, Y.: ObjectRank: Authority-based keyword search in databases. In: VLDB, pp. 564–575 (2004)
14. Hristidis, V., Hwang, H., Papakonstantinou, Y.: Authority-based keyword search in databases. ACM Trans. Database Syst. 33(1), 1–40 (2008)
15. Sayyadian, M., LeKhac, H., Doan, A., Gravano, L.: Efficient keyword search across heterogeneous relational databases. In: ICDE, pp. 346–355 (2007)

A Grid Index Based Method for Continuous Constrained Skyline Query over Data Stream

Li Zhang, Yan Jia, and Peng Zou

School of Computer, National University of Defense Technology, Changsha, China 410073
catlily1981@gmail.com,
jyjiayan@vip.sina.com,
zpeng@nudt.edu.cn

Abstract. As an essential query, skyline computation over data stream is very important for many on-line applications, including mobile environment, network monitoring, communication, sensor network and stock market trading, etc. Different from most popular skyline processing methods that deal with the whole data set, this paper focuses on constrained skyline processing over data stream. We employ a grid based index to store the tuples and put forward two algorithms to compute and maintain skyline set based on it. We also define Influence Area for every query to minimize the cells need to be processed when new tuples arrive and old tuples expire. Theoretical analysis and experimental evidences show the efficiency of proposed approaches.

Keywords: constrained skyline, data stream, grid index.

1 Introduction

As an important query over data stream, the problem of skyline computation has attracted much research attention[1~6]. Skyline query plays a great role in many applications such as network monitoring, multi-criteria decision making, communication, sensor network and stock market trading, etc. Given a data set S of d dimensions, the skyline result SK is the set of points p with $p \in S$ and there is no other point $q \in S$ can dominates p. p is said to dominate q if p is no worse than q in every single dimension but better than q in at least one dimension. The standard of *"better"* or *"worse"* varies according to different applications. For example, in a hotel book system, a tourist to the seaside wants to reserve a hotel which is both cheap and close to seashore on the internet. In this situation, cheap is *"better"*, and expensive is *"worse"*; close to the seashore is *"better"* but far away from the seashore is *"worse"*.

Since Borzonyi[5] first introduced the skyline operator into database system, there has been a lot of research works about it. Most of them focused on the whole data set. But sometimes we care about the skyline set over the part of the data set. For instance, Fig. 1 shows the skyline of cheap hotels near the beach with and without constraints. As we can see from (a), {A, B, E} is the best choice for travellers without constraint. But when travellers add a constraint, as shown in Fig. 1(b), such as "what is the best choice for me in the range of $30 to $150?" Hotel A and E will not be an answer

L. Chen et al. (Eds.): APWeb and WAIM 2009, LNCS 5731, pp. 185–197, 2009.

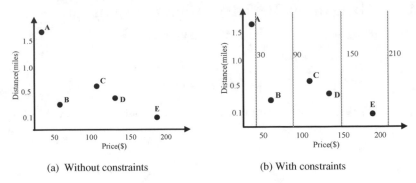

(a) Without constraints   (b) With constraints

Fig. 1. Skyline of hotels with and without constraints

because A.price is cheap than \$30, and E.price is expensive than \$150. So B is the only answer. {C, D, E} is the answer to the question "what is the best choice for me between \$90 to \$210?" Obviously, when there is a constraint (query range) attached to the skyline query, the results are different from the skyline query on the whole data set. This kind of skyline query is called constrained skyline query[1,2,11].

In data stream context, especially when the arriving rate is very high, the valid tuples are stored in main memory to satisfy real-time response. Due to the dynamic features of data stream, complexity index structure (e.g., a main memory R-tree) may be very expensive for maintaining. In this paper, we use a grid-based index instead. When a new query arrives in the system, the computation module computes its result by searching the minimum number of cells that may contain skyline tuples. The maintaining module is called when a new tuple falls in these grids or a tuple of skyline set expires.

In summary, we make the following contributions in this paper. First, we define the problem of constrained skyline query over data stream. Second, we put forward two algorithms using grid index to compute and maintain constrained skyline. Third, we conduct detailed theoretical analysis and extensive experiments to valid the effectiveness of our methods.

The rest of this paper is organized as follows. We introduce the related work in Section 2. In Section 3, the problem of constrained skyline query is formally defined. We present our algorithms in Section 4. The time and space complexity of the proposed algorithms are analysed in Section 5. Section 6 presents the results of our experimental evaluation. The conclusion is in Section 7.

2 Related Work

Skyline queries have been studied since 1960s in the theory field where skyline points are known as Pareto sets and admissible points [6] or maximal vectors [7]. Borzonyi et al. [5] first investigated the skyline computation problem in the context of databases. They put forward two algorithms BNL and D&C to process skyline query problems in database. SFS[8] improved BNL by pre-sorting the tuples. Later work proposed the index based solutions[9~11]. NN(Nearest Neighbour) method in [10] identifies skyline points by recursively invoking R-tree based depth-first nearest neighbour search. It is more efficient than the earlier algorithms and can return the results gradually. Dimitris et al.[11] put forward an efficient algorithm called *BBS* to improve the *NN* method. Like *NN*, *BBS* is also based on nearest neighbour search and use R-tree to index the entries, but it avoids the shortcomings of *NN* like visiting the overlapping regions twice. According to [11], *BBS* only processes the entries which are not dominated by current skyline tuples. *BBS* can be applied to variations of skyline query with a little modification, such as constrained skyline queries. But *BBS* is used for static data, not dynamic data like data stream. [1] and [2] consider the constrained skyline processing, but they mainly focus on parallelity among different nodes. The skyline processing over distributed environment attracts increasing attention recently, such as [12~18].

As far as the data stream context is concerned, Tao et al.[19] put forward two methods to maintain sliding window skylines over data stream. The first one is called lazy method, which delays most computational work until the expiration of a skyline point, while the eager method takes advantage of precomputation to minimize memory consumption. Lin et al.[4] focused on computing the skyline against the most recent n of N elements in a data stream set. They developed an effective pruning technique to minimize the number of elements to be kept, so the time complexity is decreased. Morse et al.[3] presented a new algorithm called LookOut for evaluating the continuous time-interval skyline efficiently. Tian[20] developed a grid index based continuous skyline processing method over data stream. All these works consider the skyline computation of the whole valid data set, which is different from our constrained skyline computation problem.

3   Preliminaries

3.1   Problem Definition

First, we will give several definitions. Suppose every tuple p of the data set S is of d dimensions, $x_1, x_2, ..., x_d$. p is represented by $p(p.x_1, p.x_2, ..., p.x_d)$. Without loss of generality, small value is preferable in this paper.

Definition 1 Dominate. For two tuples $p=(p.x_1, p.x_2, ..., p.x_d)$, $q=(q.x_1, q.x_2, ..., q.x_d)$, p dominates q (denoted as Dominate(p,q)) means for $1 \le i, j \le d$, $\forall i, p.x_i \le q.x_i$, and $\exists j$, $p x_j < q.x_j$.

For $p, q \in S$, if p does not dominate q, and q doesn't dominate p, we say p and q is not comparable, denoted by $p <> q$.

Definition 2 Skyline. Skyline is a data set which contains all points that are not dominated by any other point. Each point in the skyline set is called a skyline point. *SK* is used to denote the skyline set. As we can see $\forall p, q \in SK, p <> q$.

Defination 3 Constrained Skyline Query. A skyline query which is attached with constraints (query range) on specific dimensions is called a constrained skyline query. A constraint on a dimension is a range like (l, r) specified by the user according to their interest.

Defination 4 Key Dimension. The Dimension with a constraint specified on is called a Key Dimension.

For simplicity of expression, we consider the skyline queries with only one Key Dimension. But with a little modification, the algorithms can be used for multi Key Dimensions. We use concepts tuple and point, dimension and attribute alternately.

The problem we consider in this paper is: A set of constrained skyline queries $Q=\{Q_1(l_1, r_1), Q_2(l_2, r_2), ..., Q_m(l_m, r_m)\}$ registers in a data stream S, where the (l_i, r_i) is the constraint (query range) on Key Dimension. The sliding window is W. Each query $Q_i(l_i, r_i)$ computes its initial query results, and maintain its results according to valid tuples contained in W. Our object is to design an effective method to answer the constrained skyline query over data stream in real time.

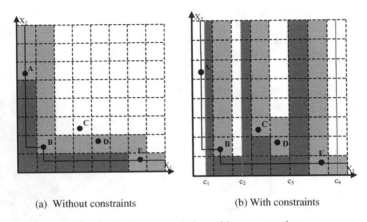

(a) Without constraints   (b) With constraints

Fig. 2. Skyline computation without constraints

3.2   Data Structure

As a popular index in data stream management, grid index is very efficient. In this paper, we adopt it to index the tuples in main memory. The extent of each cell on every dimension is δ. Assuming a 2-dimensional space, the cell $c_{i,j}$ at column i and row j contains all tuples whose $i \cdot \delta \prec p.x_1 \leq (i+1) \cdot \delta$, and $j \cdot \delta \prec p.x_2 \leq (j+1) \cdot \delta$. So given a tuple $p(p.x_1,p.x_2)$, the cell $c_{i,j}$ it belongs to can be computed by $i = \lfloor p.x_1 / \delta \rfloor$ and $j = \lfloor p.x_2 / \delta \rfloor$. The right-top point of $c_{i,j}$ is denoted by $c_{i,j}.RT$, while the left-bottom point of $c_{i,j}$ is denoted by $c_{i,j}.LB$.

Next, we will use an example to denote the problem of constrained skyline computation and the idea of our technique.

After adding a grid index to Fig. 1, as shown in Fig. 2(a), we can see that the points fall in dark cells will definitely change the skyline set, and the points fall in grey cells will possibly change the skyline set. The points in the cells without colour will not affect the skyline set at all.

Definition 5 Definite Influence Area(DIA). Definite Influence Area is a set of grids:

$\{ c_{i,j} \mid \forall p \in SK, Dominate(c_{i,j}.RT, p) \cup (c_{i,j}.RT <> p), 1 \leq i, j \leq d \}$.

Definition 6 Possible Influence Area(PIA). Possible Influence Area is a set of grids:

$$\{ c_{i,j} \mid (\forall p \in SK, \neg Dominate(p, c_{i,j}.LB)) \cap (\exists q \in SK, Dominate(q, c_{i,j}.RT)),\ 1 \le i, j \le d \}$$

The dark area in Fig. 3(a) is the DIA, while the grey area is PIA. PIA and DIA make up of Influence Area of Q. Any tuple falls into Influence Area may change the skyline result of Q.

Definition 7 Immune Area. Immune Area is a set of grids:

$$\{ c_{i,j} \mid \exists p \in SK, Dominate(p, c_{i,j}.LB),\ 1 \le i, j \le d \}$$

The cells without colour in Fig. 2(a) belong to Immune Area. The tuples fall into Immune Area will not affect the skyline set.

When a constraint imposed to a skyline query, the Influence Area and the Immune Area will change. Fig. 2(b) illustrates the Influence Area and the Immune Area of three constrained skyline queries: $Q (c_1, c_2)$, $Q(c_2, c_3)$ and $Q(c_3, c_4)$, we can see that the Influence Area are totally different from the global skyline query.

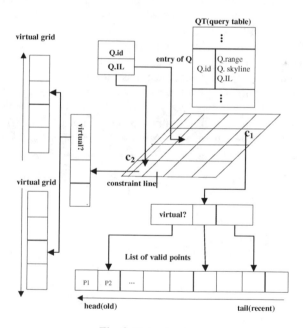

Fig. 3. Data structure

We will use the data structure listed in Fig. 3 in this paper to solve these three problems.

The running constrained skyline queries are stored in a query table QT at the right top corner of Fig. 3. QT maintains for each Q a unique identifier $Q.id$, its query range $Q.range$, its current results $Q.skyline$, and $Q.IL$. $Q.IL$ used to stores the pointers to the cells in Influence Area of the query.

In order to provide an efficient strategy for evicting expiring tuples, all the valid tuples in a cell are stored in a single list, according to their arriving time. The new tuple is placed at the end of the list, and the old ones that fall out of the window are discarded from the head of the list, which satisfies the FIFO (first-in-first-out). Each cell contains a list of pointers (PL_c) to the corresponding tuples. In fig. 3, c_2 is divided into two parts by a constraint. We call each part a virtual cell. Each virtual cell has its own PL_c. When a tuple falls into c_2, it will be examined to see which virtual part it belongs to. We assume that difference between any two constraints is bigger than or equal to δ. With this assumption, a grid can only be divided into two parts at most.

4   Constrained Skyline Maintaining Algorithm

4.1   The Computation Module

The computation module is in charge of computing the skyline set at current sliding window W. It is used in initialization. A naive way to obtain the result of a query Q_i is to compute the skyline with an existing algorithm, then compute the Influence Area of Q_i. Obviously this method is very inefficient in practice, since it needs to check all the cells twice to determine the skyline set and the Influence Area. As small value is preferable in this paper, we can see that once a skyline tuple is found in some cell, the rest cells whose left bottom point is dominated by this tuple should not be processed, as the points contained in them will not contribute to the skyline set. Based on this feature, we put forward an efficient initialization algorithm CS_CM without having to process all the cells. The pseudocode is shown in Fig. 4. Suppose all attribute values range from 0 to 1, so the number of cells in one dimension is $1/\delta$. For each $Q_i(l_i, r_i)$, the left bottom cell is $c_{\lfloor l_i/\delta \rfloor,0}$, the left top bottom cell is $c_{\lfloor l_i/\delta \rfloor,1/\delta}$, the right bottom cell is $c_{\lfloor r_i/\delta \rfloor,0}$.

CS_CM first compute the Max_{x1}, Max_{x2}(the max value in x_1 and x_2 axis), and the left bottom cell $c_{i,j}$ of Q. Then initialize an empty skyline set SK and an empty heap H. Starting from $c_{i,j}$, CS_CM computes the skyline set of Q. If there are skyline tuples in $c_{i,j}$, there is no need to process the cell whose subscripts are bigger than i and j, such as $c_{i+1,j+1}$. Set the boolean variable $continue$ to 0, meaning that the algorithm can end after this loop is finished. Otherwise, insert the skyline tuples into SK.

No matter there is or not a skyline tuple in $c_{i,j}$, add a pointer in $Q.IL$ pointing to $c_{i,j}$. There are may be skyline tuples in $(c_{i+1,j}, c_{i+2,j}, ..., c_{max\ x1,j})$ and $(c_{i,j+1}, c_{i,j+2,...,} c_{i,maxx2})$, since their LB is not dominated by the skyline point in $c_{i,j}$. Put these cells into H and process them. If the left bottom point of next cell is dominated by one point in current skyline set or H is empty, the second while loop (line 16~20) ends. Otherwise, if there is skyline tuple in a cell, insert the tuple into SK and add a pointer in $Q.IL$ pointing to the cell. ADD 1 to i and j separately. CS_CM ends when any of these three conditions is satisfied: $continue$ is 0 or $i= Max_{x1}$, or $j= Max_{x2}$. Fig. 5 illustrates how CS_CM compute the skyline set for Q without visiting all the cells.

```
Algorithm: CS_CM
Input: Data set S,  constrained skyline query  Q(l, r)
Output: SK (Skyline of P) and Q.IL (Initial influence region)
Max_x1 = ⌊r_i / δ⌋, Max_x2 =1/ δ ;                                        1
i = ⌊l_i / δ⌋; j=0;                                                       2
Initialize skyline set SK, a max-heap H, a list Q.IL for Q.              3
Boolean continue =1;                                                     4
While (continue =1& c_{i,j}.LB is not dominated by any point in SK&
i<= Max_x1 &j<= Max_x2)                                                  5
{                                                                        6
    Compute skyline set of c_{i,j} and insert the skyline tuple into SK;  7
    If there is a skyline tuple in c_{i,j}                               8
        continue=0;                                                      9
    m = i; n = j;                                                        10
    Put c_{m,n} into Q.IL;                                              11
    For (; m<= Max_x1;)                                                 12
        If c_{++m, n} is not dominated by any tuple in SK               13
            Put c_{m, n} into H;                                        14
    For (;n<= Max_x2;)                                                  15
        If c_{m,++n} is not dominated by any tuple in SK                16
            Put c_{m,n} into H;                                         17
    While(H is not empty)                                              18
    {                                                                  19
        Compute skyline set in the cell and insert into SK;            20
        Put the cell into Q.IL;                                        21
    }                                                                  22
    i++; j++;                                                          23
}                                                                      24
```

Fig. 4. The pseudocode of *CS_CM*

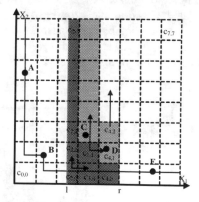

Fig. 5. Process of computing skyline set for Q(l, r)

4.2 Maintenance Module

In this part, we will detail our constrained skyline maintenance module *CS_MM*.

There only two situations when *SK* needs to be updated. The first one is when a new tuple arrives at system. The second one is when an old skyline tuple expires. The pseudocode is listed in Fig. 6.

Algorithm: *CS_MM*
Input: new arriving tuple p or an expiring tuple q
Output: updated *SK* and *Q.IL* (If needed)

// handle insertion

For every new arrival tuple p	1
Add p into the corresponding $c_{i,j}.PL$;	2
For every Q(l,r) which has a pointer to $c_{i,j}$	3
Compare p with every tuple in SK	4
If p is not dominated by any tuple in SK	5
Insert p into the SK;	6
Remove tuples in SK which are dominated by p;	7
Update Q.IL, Remove the pointers to the cells in Q.IL which are dominated by p;	8
// handle the expiration	9
When an old tuple q in $c_{i,j}$ expires	10
Delete it from the $c_{i,j}.PL$;	11
For every Q(l, r) whose Q.SK contains q	12
Delete q from Q.SK;	13
Repeat line 4~24 in CS_CM to compute new skyline tuples which are only dominated by q;	14

Fig. 6. Pseudocode of CS_*MM*

Returning to the example in Fig. 5, assume that tuple F and G arrive at the system. F is processed first. Since F dominates C and D, F is inserted into the *Q.SK*, C and D is deleted from *Q.SK*. G is inserted into $c_{4,2}.PL$. The cells of Influence Area of Q like $(c_{3,1}, c_{3,2},..., c_{3,7})$ and $c_{4,1}, c_{4,2}$ are dominated by F, so they are deleted from the Influence Area of Q. *Q.IL* is updated. G is dominated by F, so just insert G into $c_{4,2}.PL$. *SK* and *Q.IL* are not affected. Invalid skyline points appear hollow in Fig 7.

Assume at the next moment, K arrives at the cell $c_{3,0}$, and C, D, F, G expire. Insertion is processed first. Since K is not dominated by F, K is inserted into *Q.SK*. The

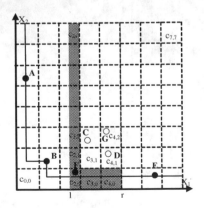

Fig. 7. F and G arrive

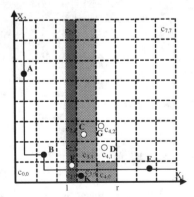

Fig. 8. K arrives and C, D, F, G expire

subscript of the cell whose left bottom point is dominated by K starts from (4, 1). Because $c_{4,1}$ doesn't belong to Influence Area, so $Q.IL$ remains same. Then C, D, F, G expires. C, D, and G are not skyline tuples, so they are just deleted from the PL of corresponding cell. F belongs to the $Q.SK$. Delete F from $Q.SK$ and start from $c_{2,0}$ looking for the new skyline tuples, the idea is the same with the one we used in CS_CM. The procedure is shown in Fig. 8.

5   Complexity Analysis

In accord with previous work, we make two assumptions: (1) the average data cardinality at each timestamp is N, (2) the tuples are uniformly distributed in a unit d-dimensional workspace and the average arriving rate is r. As we mentioned before, δ is the cell extent per axis. Suppose all attribute values range from 0 to 1, the number of cells in one dimension is $1/\delta$. The total number of cells is $(1/\delta)^d$ and each cell contains $N \cdot \delta^d$ tuples on average. A cell is denoted by $c_{i_1, i_2, ..., i_n}$, where i_j ranges from 1 to $1/\delta$.

5.1   Time Complexity

We analyse the running time of CS_CM first. Because the tuples are uniformly distributed, so there are tuples in the left bottom cell of a query Q. Without loss of generality, suppose the key dimension is the first dimension. The Influence Area of Q is a set of cells : $IA=\{ c_{i_1, i_2, ..., i_d} | \lfloor l/\delta \rfloor \leqslant i_l \leqslant \lfloor r/\delta \rfloor$ & for $1<j<=d$, at least one i_j equals to 0$\}$. $|IA|$ equals to the number of pointers in $Q.IL$, which means $|IA|$ equals to $|IL|$ for Q. So there are $|IA|$ cells need to be processed. The rest are dominated by some tuple in the left bottom cell. Suppose the cost to process a cell on average is C, so the time complexity of CS_CM is $O(C \cdot |IA|)$.

Concerning the cost of CS_MM, in every processing cycle, r new tuples arrive at the system, while r old tuples expire. Hence, the cell update time is $O(r)$. Each cell receives $r \cdot \delta^d$ insertions and $r \cdot \delta^d$ deletions. For the Influence Area, it will cost $O(|IA| \cdot r \cdot \delta^d \cdot |SK|)$ to check if SK should change. The probability of a new tuple becomes a skyline result can be approximated by $\frac{|SK|}{N}$. When a new tuple becomes a skyline result, the cells in Influence Area need to be examined to see if they are dominated by the new skyline tuple. That will cost $O(|IA|)$ at most. For any expiring tuple q belonging to $Q.|SK|$, we need to find a new skyline tuple that is dominated only by q. In the worst case when q is in the left bottom cell, the cost will be the same as CS_CM, $O(C \cdot |IA|)$. So the total cost of CS_MM at the worst case will be

$$O(|IA| \cdot r \cdot \delta^d \cdot |SK| + r \cdot \frac{|SK|}{N}(|IA| + C \cdot |IA|)).$$

5.2 Space Complexity

In this paper, $N \cdot d$ memory units are need to store N valid d-dimensional tuples, and N memory units to store the pointers in the point lists of the cells. For every $Q_i(i=1,3,...,m)$, we need 1 memory unit for its id, 2 for its range, $|SK_i|$ for its results, and $|IA_i|$ for its influence list. In conclusion, the space requirements of our algorithms are $O \left((N + 1) \cdot d + \sum_{1}^{m} (|SK_i| + |IA_i|) \right)$.

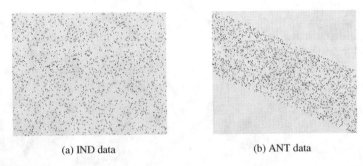

(a) IND data   (b) ANT data

Fig. 9. Two data set of 2-dimension

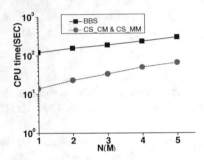

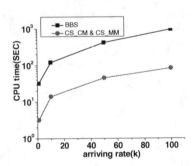

Fig. 10. The effect of cardinality   **Fig. 11.** The effect of arrived rate

6 Experimental Evaluation

In this section, we experimentally evaluate our algorithms using data sets of independent (IND) and anti-correlated (ANT). The dimensionality d ranges from 2 to 6. ANT and IND shown in Fig. 9 are very popular synthetic data set. For IND data, the attribute values of each tuple are generated independently, following a uniform distribution, while the values of ANT data increase with time. All the experiments are conducted on an AMD Athlon 64 PC with 2.09 GHz processor, 1GB main memory, and Windows XP OS. We implement a simulation system programmed in C++ to validate the performance of the *CS_CM& CS_MM*.

   Our experiments use a count-based window with size N ranging from 1 million to 5 million and the rate of data stream r is 10,000/s. The query range (l_i, r_i) are generated randomly. We conduct three experiments in this section, studying the performance of our skyline algorithm *CS_CM & CS_MM*. Since *BBS* can process constrained skyline queries, we will compare with *BBS*. For all experiments, we store data structures in memory to simulate continuous skylines queries of streaming applications.

6.1 The Effect of Cardinality and Arriving Rate

First, we show the effect of cardinality N and arriving rate r on *BBS* and *CS_CM & CS_MM*. we use the IND data set. The dimension d is 4. The number of query in Fig. 10 is 1K. The N in Fig. 11 is 1M. We can see from these two figures that our algorithms *CS_CM & CS_MM* perform better than *BBS*. The reason is that *CS_MM* is very efficient with each new insertion or deletion, but *BBS* has to recompute the skyline results from the scratch when the new tuple arrives and old tuple expires. We also can see that when the arriving rate is very high, *BBS* is not applicable to handle the data stream environment.

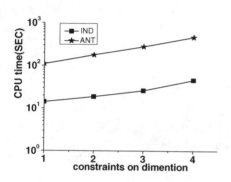

Fig. 12. The effect of query number   **Fig. 13.** The effect of dimension

6.2 The Effect of Constraints

In this experiment, we study how the number of constraints affects the algorithms. We add constraints to 1, 2, 3, 4 dimensions orderly to see the effect. We use IND and ANT data set separately, and the dimension d is 4. Fig. 12 shows the experiment result.

As constraints are added on more dimensions, the CPU time increases too, because cells are partitioned by the more constraints, which brings more expenses to the system for maintaining the data structure. The IND data set performs much better because the tuples are distributed in every cell uniformly. So the skyline tuples are mostly found in the first recurrence. But for the ANT data set, more cells needs to be processed.

6.3 The Effect of Dimension

Like *BBS*, the performance of *CS_CM & CS_MM* degrades with the increase of dimensionality. As the number of attributes increases, more cells need to maintain in the system, resulting in more cells to be processed for computing skyline tuples. The maintaining of Influence Area will be more complicated too. Our experiment shown in Fig.13 proved this.

7 Conclusion

In this paper, we study the problem of constrained skyline processing over data stream. Effective algorithms *CS_CM* and *CS_MM* are presented for online constrained skyline computation and maintaining. We use a grid based index and define the Influence Area for every constrained query, which minimizes the tuples needs to be processed. The experiments and theoretical analysis show the effectiveness of our algorithms.

Acknowledgment. This work is supported by the National High-Tech Research and Development Plan of China ("863" plan) under Grant No. 2006AA01Z451, 2007AA010502 and 2007AA01Z474. The authors would like to thank the anonymous reviewers for their constructive comments.

References

1. Cui, B., Lu, H., Xu, Q., et al.: Parallel Distributed Processing of Constrained Skyline Queries by Filtering. In: Proc. ICDE (2008)
2. Wu, P., Zhang, C., Feng, Y., Zhao, B.Y., Agrawal, D.P., El Abbadi, A.: Parallelizing skyline queries for scalable distribution. In: Ioannidis, Y., Scholl, M.H., Schmidt, J.W., Matthes, F., Hatzopoulos, M., Böhm, K., Kemper, A., Grust, T., Böhm, C. (eds.) EDBT 2006. LNCS, vol. 3896, pp. 112–130. Springer, Heidelberg (2006)
3. Morse, M., Patel, J.M., Grosky, W.I.: Efficient Continuous Skyline Computation. In: Proc. ICDE (2006)
4. Lin, X., Yuan, Y., Wang, W., et al.: Stabbing the Sky: Efficient Skyline Computation over Sliding Windows. In: Proc. ICDE (2005)
5. Borzsonyi, S., Kossmann, D., Stocker, K.: The Skyline Operator. In: Proc. ICDE (2001)
6. Barndor-Nielsen, O., et al.: On the distribution of the number of admissable points in a vector random sample. Theory of Probability and its Application 11(2) (1966)
7. Bentley, J.L., et al.: On the average number of maxima in a set of vectors and applications. Journal of ACM 25(4) (1978)
8. Chomicki, J., Godfrey, P., Gryz, J., et al.: Skyline with Presorting. In: Proc. ICDE (2003)
9. Tan, K., Eng, P., Ooi, B.-C.: Efficient progressive skyline computation. In: Proc. VLDB (2001)
10. Kossmann, D., Ramsak, F., Rost, S.: Shooting stars in the sky: An online algorithm for skyline queries. In: Proc. VLDB (2002)
11. Papadias, D., Tao, Y., Fu, G., et al.: Progressive skyline computation in database systems. ACM Transactions on Database Systems 30 (2005)
12. Pei, J., Jin, W., Ester, M., et al.: Catching the best views of skyline: A semantic approach based on decisive subspaces. In: Proc. VLDB (2005)
13. Yuan, Y., Lin, X., Liu, Q., et al.: Efficient computation of the skyline cube. In: Proc. VLDB (2005)

14. Vlachou, A., Doulkeridis, C., Kotidis, Y., et al.: SKYPEER: Efficient Subspace Skyline Computation over Distributed Data. In: Proc. ICDE (2007)
15. Balke, W.-T., Güntzer, U., Zheng, J.X.: Efficient distributed skylining for web information systems. In: Bertino, E., Christodoulakis, S., Plexousakis, D., Christophides, V., Koubarakis, M., Böhm, K., Ferrari, E. (eds.) EDBT 2004. LNCS, vol. 2992, pp. 256–273. Springer, Heidelberg (2004)
16. Wang, S., Ooi, B.C., Tung, A.K.H., et al.: Efficient skyline query processing on peer-to-peer networks. In: Proc. ICDE (2007)
17. Jagadish, H.V., Ooi, B.C., Vu, Q.H.: BATON: A balanced tree structure for peer-to-peer networks. In: Proc. VLDB (2005)
18. Huang, Z., Lu, C.S.J.H., Ooi, B.C.: Skyline queries against mobile lightweight devices in MANETs. In: Proc. ICDE (2006)
19. Tao, Y., Papadias, D.: Maintaining Sliding Window Skylines on Data Streams. IEEE Transactions on Knowledge and Data Engineering 18 (2006)
20. Tian, L., Zou, P., Li, A., Jia, Y.: Grid Index Based Algorithm for Continuous Skyline Computation. Chinese Journal of Computers 31 (2008)

Ontology Personalization: An Approach Based on Conceptual Prototypicality

Xavier Aimé[1,3], Frédéric Furst[2], Pascale Kuntz[1], and Francky Trichet[1]

[1] LINA - Laboratoire d'Informatique de Nantes Atlantique (UMR-CNRS 6241)
University of Nantes - Team "Knowledge and Decision"
2 rue de la Houssinière BP 92208 - 44322 Nantes Cedex 03 - France
pascale.kuntz@univ-nantes.fr, francky.trichet@univ-nantes.fr
[2] MIS - Laboratoire Modélisation, Information et Système
University of Amiens
UPJV, 33 rue Saint Leu - 80039 Amiens Cedex 01 - France
frederic.furst@u-picardie.fr
[3] Société TENNAXIA
37 rue de Châteaudun - 75009 Paris - France
xaime@tennaxia.com

Abstract. With the current emergence of Cognitive Sciences and the development of Knowledge Management applications in Social and Human Sciences, *Subjective Knowledge* becomes an unavoidable subject and a real challenge, which must be integrated and developed in Ontology Engineering and Ontology-based Information Retrieval. This paper introduces a new approach dedicated to the Personalization of a Domain Ontology. Inspired by works in Cognitive Psychology, our work is based on a process which aims at capturing the user-sensitive degree of truth of the *categorisation process*, that is the one which is really perceived by the end-user. Practically, this process consists in decorating the Specialisation/Generalisation links (*i.e.* the *ISA* links) of the hierarchy of concepts with a specific gradient. As this gradient is defined according to the three aspects of the semiotic triangle (*i.e.* intensional, extensional and expressional dimension), we call it **Semiotic-based Prototypicality Gradient**. It enrichs the initial formal semantics of an ontology by adding a pragmatics defined according to a context of use which depends on parameters like culture, educational background and/or emotional context of the end-user.

Keywords: Contextual Ontology, Typicality, Categorisation, Conceptual prototypicality, Semiotic measure, Information Retrieval, Personalisation, Semantic Web, Pragmatic Web.

1 Introduction

This paper deals with knowledge which is included in the semantic and episodic memory of Human Being [7]. This knowledge, which can be expressed through textual, graphic or sound documents, corresponds to what must be captured

L. Chen et al. (Eds.): APWeb and WAIM 2009, LNCS 5731, pp. 198–209, 2009.

within a Domain Ontology, as it is specified by the consensual definition of T. Gruber [6]: *"an ontology is a formal and explicit specification of a shared conceptualisation"*. The advent of the Semantic Web and the standardisation of a Web Ontology Language (OWL) have led to the definition and the sharing of a lot of ontologies dedicated to scientific or technical fields. Our work aims at providing measures dedicated to the personalization of a Domain Ontology. This personalization process mainly consists in adapting the content of an ontology to its context of use. Our approach of ontology personalization aims at taking these parameters into account in order to reflect the degree of truth users of ontologies perceive on the *is-a* hierarchies and to what extent the terms associated to the concepts are representative. According to the model of Pierce [12], any perceptible phenomenon is perceived according to three dimensions: syntax, semantics and pragmatics. Our work is based on the semiotic triangle, as defined by the linguists Ogden and Richard [10]. The three corners of this triangle are (1) the *reference* (*i.e.* the intensional dimension) which is a unit of thought defined from abstraction of properties common to a set of objects (*i.e.* the concepts of an ontology and their properties), (2) the *referent* (*i.e.* the extensional dimension) which corresponds to any part of the perceivable or conceivable world (*i.e.* the instances of concepts) and (3) the *term* (*i.e.* the expressional dimension) which is a designation of an unit of thought in a specific language (*i.e.* the linguistic expressions used to denote the concepts). The goal of our gradient, called **Semiotic-based Prototypicality Gradient**, is to capture the user-sensitive degree of truth of the *categorisation process*, that is the one which is perceived by the end-user.

The rest of this paper is structured as follows. Section 2 introduces the formal definition of the Semiotic-based Prototypicality Gradient. Section 3 presents the distributional analysis of the SPG and experimental results.

2 Semiotic-Based Prototypicality Gradient (SPG)

Defining an ontology O of a domain D at a precise time T consists in establishing a consensual synthesis of individual knowledge belonging to a specific endogroup; an endogroup is a set of individuals which share the same distinctive signs and, therefore, identify a community. For the same domain, several ontologies can be defined by different endogroups. We call *Vernacular Domain Ontologies* (VDO) this kind of resources[1]. This property is also described by E. Rosch as *ecological* [4,14], in the sense that although an ontology belongs to an endogroup, it also depends on the context in which it evolves. Thus, given a domain D, an endogroup G and a time T, a VDO depends on three factors, characterising a precise context: (1) the culture of G, (2) the educational background of G and (3) the emotional state of G. In this way, a VDO can be associated

---

[1] *Vernacular*, which comes from the latin word *vernaculus*, means native. For instance, vernacular architecture, which is based on methods of building which use locally available resources to address local needs, tends to evolve over time to reflect the environmental, cultural and historical context in which it exists.

to a pragmatic dimension. Indeed, a same VDO can be viewed (and used) from multiple points of view, where each point of view, although not reconsidering the formal semantics of D, allows us to adapt (1) the degrees of truth of the *isa* links defined between concepts and (2) the degrees of expressivity of the terms used to denote the concepts. We call *Personalised Vernacular Domain Ontologies* (PVDO) this kind of resources. Our work is based on the fundamental idea that all the sub-concepts of a decomposition are not *equidistant* members, and that some sub-concepts are more representative of the super-concept than others. This phenomenon is also applicable to the set of terms used to denote a concept. This assumption is validated by works in Cognitive Psychology [7,8]. Formally, our gradient is based on a Vernacular Domain Ontology (VDO), given a field D and an endogroup G. This type of ontology is defined by the t-uple $O_{(D,G)} = \{\mathcal{C}, \mathcal{P}, \mathcal{I}, \Omega_{(D,G)}, \leq^{C}, \sigma_P, L\}$ where:

- $\mathcal{C}, \mathcal{P}, \mathcal{I}$ represent respectively the disjoined sets of concepts, properties[2] and instances;
- $\Omega_{(D,G)}$ is a set of documents (*e.g.* text, graphic or sound documents) related to a domain D and shared by the members of the endogroup G;
- $\leq^{\mathcal{C}}$: $\mathcal{C} \times \mathcal{C}$ is a partial order on $\mathcal{C}$ defining the hierarchy of concepts ($\leq^{C} (c_1, c_2)$ means that the concept c_1 subsumes the concept c_2);
- $\sigma_P : \mathcal{P} \to \mathcal{C} \times \mathcal{C}$ defines the domain and the range of a property;
- $L = \{L_C, f_{term_C}\}$ is the lexicon related to the dialect of G where (i) L_C represents the set of terms associated to $\mathcal{C}$, (ii) the function $f_{term_c} : \mathcal{C} \to (L_C)^n$ which returns the tuple of terms used to denote a concept.

We define $spg_{G,D} : \mathcal{C} \times \mathcal{C} \to [0,1]$ the function which, for all couple of concepts $c_f, c_p \in \mathcal{C}$ such as it exists an *is-a* link between the super-concept c_p and the sub-concept c_f, returns a real (null or positive value) which represents the conceptual prototypicality gradient of this link, in the context of a PVDO dedicated to a domain D and an endogroup G. For two concepts c_p and c_f, this function is formally defined as follows:

$$spg_{G,D}(c_p, c_f) = [\alpha * intent(c_f, c_p) + \beta * expres_{G,D}(c_f, c_p) + \gamma * extens_{G,D}(c_f, c_p)]^{\delta}$$

with (1) $\alpha + \beta + \gamma = 1$, where $\alpha \geq 0$ a weighting of the intensional component, $\beta \geq 0$ a weighting of the expressional component, $\gamma \geq 0$ a weighting of the extensional component, and (2) $\delta \geq 0$ a weighting of the mental state of the endogroup G. The main advantage of our approach is that (1) it integrates the intensional, extensional and expressional dimensions of a conceptualisation for defining how a sub-concept is *representative/typical* of its super-concept and the influence of these dimensions can be modulated via the α, β and γ parameters and (2) it allows us to modulate this representativeness according to an emotional dimension via the δ parameter. The values of α, β and γ are defined (manually) according to the context of the ontology personalization process. Indeed, when no instances (or few) are associated to the ontology then it is relevant to minimize

---

[2] Properties include both attributes of concepts and domain relations.

the influence of the extensional dimension by assigning a low value to γ. In a similar idea, when the ontology does not include properties then it is relevant to minimize the influence of the intensional dimension by assigning a high value to α. And when the ontology is associated to a huge and rich textual corpora then it is relevant to maximize the influence of the expressional dimension by assigning a high value to β, knowing that $\alpha + \beta + \gamma = 1$. The value of δ is used to modulate the influence of the emotion state on the perception of the conceptualisation. Multiple works on the influence of emotions on human evaluation have been done in psychology [3,11]. The conclusion of these works can be summarized as follows: when we are in a negative mental state (*e.g.* fear or nervous breakdown), we tend to centre us on what appears to be the more important from an emotional point of view. In the context of our approach, it consists in reducing the universe to what is very familiar; for instance, our personal dog (or the one of a neighbor) - which at the beginning is inevitably the most characteristic of the category - becomes **the** and quasi unique dog. Respectively, in a positive mental state (*e.g.* love or joy), we are more open in our judgment and we accept more easily the elements which are not yet be considered as so characteristic. According to [9], a negative mental state leads to the reduction of the value of representation, and conversely for a positive mental state. Thus, we characterize: (1) a *negative* mental state by a value $\delta \in]1, +\infty[$, (2) a *positive* mental state by a value $\delta \in]0, 1[$, and (3) a *neutral* mental state by the value 1. When the value of δ is low, the value of the gradients associated to the concepts which are initially not considered as being so representative increases considerably, because a positive state facilitates the open mind, the valorisation, etc. Conversely, when the value of δ is high (*i.e.* a strongly negative mental state), the effect is to *select* only the concepts which own a high value of typicality, eliminating *de facto* the other concepts.

2.1 Intensional Component

The intensional component of our gradient aims at taking (i) the structure of a conceptualisation and (ii) the definition in intension of its components into account. In order to compare two concepts from an intensional point of view, we propose a measure based on the properties shared by the sub-concepts as developed in [1,2]. For each concept $c \in \mathcal{C}$, we define a *Characteristic Vector* (CV) $\vec{v_c} = (v_{c1}, v_{c2}, ..., v_{cn})$ with $n = |\mathcal{P}|$, and $v_{ci} \in [0,1], \forall i \in [1,n]$ a weight assigned to each property p_i of $\mathcal{P}$. A concept is defined by the union of all the properties whose weights are not null. The set of concepts corresponds to a point cloud defined in a space with $|\mathcal{P}|$ dimensions. When assigning weights to properties, one has to respect a constraint related to the ISA relationship: a concept c is subsumed by a concept d (noted $\leq^C (d,c)$) if and only if $v_{ci} \geq v_{di}, \forall i \in [1,n]$ with $n = |\mathcal{P}|$. For any $c \in \mathcal{C}$, we define a prototype concept from all the sub-concepts of c. This prototype concept is characterized by a *Prototype Vector* (PV) $\vec{t_c} = (t_{c1}, t_{c2}, ..., t_{cn})$, with $n = |\mathcal{P}|$. Prototype concepts correspond to summaries of semantic features characterizing categories of concepts. They are stored in the episodic memory, and are used in the process of categorization per comparison. In our work, we consider that a prototype concept of a concept c corresponds to the barycenter of the point cloud formed by the set of the CV of all concepts belonging to the descent of c^3. Thus, the *Prototype Vector* of a concept c is formally defined as follows:

$$\vec{t_c} = \frac{1}{\sum_{s \in S} \lambda(s)} \sum_{s \in S} \lambda(s) \, \vec{v_s}$$

Where:

- $\lambda(s)$ is equal to $\frac{depth_{tree}(c) - depth(s) + 1}{depth_{tree}(c)}$ with :
  - $depth_{tree}(c)$, the depth of the sub-tree having for root c;
  - $depth(s)$, the depth of s in the sub-tree having for root c.
- S, the set of concepts belonging to the descent of c.

The objective of the coefficient $\lambda(s)$ (for a concept s) is to relativize the properties which are hierarchically distant from the super-concept (cf. the use of the ratio of depths)[4]. In our work, we advocate the following principle: the more a concept is close to the prototype concept, the more it is representative of its super-concept. We consider this value as being the Euclidean normalized distance between (i) the PV of the super-concept and (ii) the CV of the sub-concept which is considered; it corresponds to the normalized distance between a point and the barycenter of the point cloud. The function *intent* : $\mathcal{C} \times \mathcal{C} \rightarrow [0,1]$ is formally defined as follows:

$$intent(c_f, c_p) = 1 - dist(\vec{t_{cp}}, \vec{v_{cf}})$$

The more the value of this function is near to 1, the more the concept c_f is *representative/typical* of the concept c_p, from an intensional point of view.

In order to illustrate and to clarify the calculation of this component, let us consider a simple tree-based hierarchy of concepts (*cf.* figure 1). This hierarchy is composed of (i) a root concept x_0, (ii) two concepts (x_1 and x_2) for the first level, and (iii) four concepts for the second level (x_{ij} such as $\leq^C (x_i, x_{ij})$). Each concept inherits the properties of its super-concept, to which it adds his own properties (starting from a set composed of 10 properties). A weight is associated to each property p in the context of each concept c. This weight evaluates how the property p is important for defining the concept c. For instance, the weight of p_1 for the concept x_1 is 0.75; for the concept x_2, it is 0.65. The first step consists in calculating the PV of x_0, from all the concepts of its descent

---

[3] We understand by descent all the sub-concepts of c, from generation 1 to n (*i.e.* the leaves).

[4] Contrary to [2], we propose to extend the calculation of the prototype to all the descent of a concept, and not only to its direct sub-concepts (*i.e.* only one level of hierarchy). Indeed, we think that all the concepts belonging to the descent (and in particular the leaves) contribute to the definition of the prototype from a cognitive point of view.

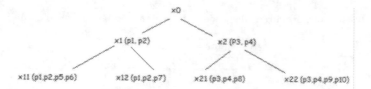

Fig. 1. Intuitive example

Table 1. Weighting of properties

cpt	x_1	x_2	x_{11}	x_{12}	x_{21}	x_{22}
p_1	0.75	0	0.75	0.86	0	0
p_2	0.65	0	0.72	0.66	0	0
p_3	0	0.45	0	0	0.56	0.46
p_4	0	0.25	0	0	0.45	0.33
p_5	0	0	0.12	0	0	0
p_6	0	0	0.66	0	0	0
p_7	0	0	0	0.81	0	0
p_8	0	0	0	0	0.45	0
p_9	0	0	0	0	0	0.55
p_{10}	0	0	0	0	0	0.72

(*i.e.* from x_1 to x_{22}). For the dimension p_1, the value of this PV is equal to the sum of:

$\frac{2-1+1}{2} * 0.75 + \frac{2-1+1}{2} * 0$ from x_1 and x_2 (equal to 0.75)

and $\frac{2-2+1}{2} * 0.75 + \frac{2-2+1}{2} * 0.86$ from x_{11} and x_{12} (equal to 0.81)

and $\frac{2-2+1}{2} * 0 + \frac{2-2+1}{2} * 0$ from x_{21} and x_{22} (equal to 0)

And all weighting by:
$\frac{1}{\frac{2-1+1}{2} + \frac{2-1+1}{2} + \frac{2-2+1}{2} + \frac{2-2+1}{2} + \frac{2-2+1}{2} + \frac{2-2+1}{2}}$ (equal to 0.25).

In this case, the coordinate of the PV for the dimension p_1 is 0.39. At the end of this process, we obtain the following PV (defined in a 10 dimensional space):
$\vec{v_{x_0}} = (0.39, 0.34, 0.24, 0.16, 0.02, 0.08, 0.1, 0.06, 0.07, 0.09)$.
The second step consists in calculating the Euclidean normalized distance between each sub-concept (defined by its Characteristic Vector) and the prototype concept of x_0 defined by the Prototype Vector v_{x_0} (*cf.* table 2). This operation is done starting from the normalized vectors of each concept, *i.e.* the coordinates of each vector divided by their length. We only calculate the intensional component of our gradient in the context of a super-concept (here x_0) and one of its direct sub-concepts (here x_1 or x_2). This component (equal to 1 - distance) has the value 0.43 for the couple (x_0, x_1) and 0.11 for the couple (x_0, x_2). Thus, from

Table 2. Euclidean normalized distance between each sub-concept and the prototype

-	x_1	x_2	x_{11}	x_{12}	x_{21}	x_{22}
$d(x_0, x_i)$	0.57	0.89	0.62	0.66	0.89	0.98

an intensional point of view, the concept x_1 is more prototypic of the concept x_0 than the concept x_2.

2.2 Expressional Component

The expressional component of our gradient aims at taking the expressional view of a conceptualisation into account, through the terms used to denote the concepts. This approach is based on the appearance frequency of a concept related to a domain D, in a universe of the endogroup G. In this way, the more an element is frequent in the universe, the more it is considered as *representative/typical* of its category. This notion of typicality is introduced in the work of E. Rosch [4,14]. In our context, the universe of an endogroup is composed of the set of documents identified by $\Omega_{(D,G)}$. Our approach is inspired by the idea of Information Content introduced by Resnik [13]. Indeed, this is not because an idea is often expressed that it is really true and objective. Psychologically, it is recognised that the more an event is presented (in a frequent way), the more it is *judged* probable without being really true for an individual or an endogroup; this is one of the ideas defended by A. Tversky in its work on the evaluation of uncertainty [15]. The function $expres_{G,D}(c_f, c_p) : \mathcal{C} \times \mathcal{C} \rightarrow [0, 1]$ is formally defined as follows[5]:

$$expres_{G,D}(c_f, c_p) = \frac{Info(c_f)}{Info(c_p)}$$

where:

$$Info(c) = \sum_{term \in world(c)} \left(\frac{count(term)}{N} * \frac{count(doc, term)}{count(doc)} \right)$$

with:

- $Info(c)$ defines the information content of the concept c;
- $count(term)$ returns the weighting number of *term* occurrences in the documents of $\Omega_{(D,G)}$. Note that this function takes the structure of the documents into account. Indeed, in the context of a scientific article, an occurence of a term t located in the keywords section is more important than another occurence of t located in the summary or the body of text. Thus, this function is formally defined as follows:

---

[5] This function is only applicable if it exists:

- a direct *is-a* link between the super-concept c_p and the sub-concept r_f, with an order relation $c_f \leq c_p$,
- or an indirect link composed of a serie of *is-a* links between the c_p and c_f.

$$count(term) = \sum_{i=1}^{m} M_{term,i}$$

where $M_{term,i} \in Z$ is the hierarchical coefficient relating to the position (in the structure of the document) of the ith occurrence of the term. The values of these coefficients are fixed in a manual and consensual way by the members of the endogroup.

- $count(doc, term)$ returns the number of documents of $\Omega_{(D,G)}$ where the $term$ appears;
- $count(doc)$ returns the number of documents of $\Omega_{(D,G)}$;
- $world(c)$ returns all the terms concerning the concept c via the function f_{term_c} and all its sub-concepts from generation 1 to generation n;
- N is the sum of all the weighting numbers of occurrence of all the terms contained in $\Omega_{(D,G)}$.

Intuitively, the function $Info(c)$ allows us to calculate "the ratio of use" of a concept in an universe, by using first the terms directly associated to the concept and then, by using the terms associated to all its sub-concepts, from generation 1 to generation n. We balance each frequency by the ratio between the number of documents where the term is present and the global number of documents. An idea which is frequently presented in few documents is less relevant than an idea which is perhaps less defended in each document but which is presented in a lot of documents of the endogroup's universe.

2.3   Extensional Component

The extensional component of our gradient aims at taking the extensional view of a conceptualisation into account, through the instances. This approach is based on the quantity of instances of a concept related to a domain D, in an universe of the endogroup G. In this way, the more a concept is frequent in the universe (because it owns a lot of instances), the more it is considered as *representative/typical* of its category. The function $extens_{G,D}(c_f, c_p) : C \times C \rightarrow [0, 1]$ is formally defined as follows:

$$extens_{G,D}(c_f, c_p) = 1 / \left(1 - \log\left(\frac{count_I(c_f)}{count_I(c_p)}\right)\right)$$

Where the function $count_I(c) : C \times I \rightarrow Z$ return the number of instances $i \in I$ of a concept $c \in C$. The form $\frac{1}{1-\log(x)}$ has been adopted in order to obtain a non-linear behavior which is more close to human judgment.

3   Experimental Results

3.1   Distributional Analysis of the Gradient

In order to evaluate the distributional analysis of the SPG values on different types of hierarchies of concepts, we have developed a specific prototype whose parameters (given an ontology O) are: N the number of concepts of O, H the depth of O, and W the max width of O. From these parameters, the prototype

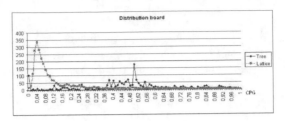

Fig. 2. Influence of the number of edges (with a constant number of concepts)

automatically generates a random hierarchy of concepts. The results presented in figure 2 have been calculated in the following context: (1) a hierarchy O_1 based on a tree described by (N=800, H=9, W= 100), (2) a hierarchy O_2 based on a lattice with a density of 0.5 described by (N=800, H=9, W= 100); and (3) $\alpha = 0.3$, $\beta = 0.3$, $\gamma = 0.3$ and $\delta = 1$. These results clearly attest the fact that multiple inheritance leads to a dilution of the typicality notion.

The results presented in figure 3 have been calculated in the following context: (1) a hierarchy O_1 based on a tree described by (N=800, H=9, W= 100); (2) a hierarchy O_2 based on a tree described by (N=50, H=2, W= 30); and (3) $\alpha = 0.3$, $\beta = 0.3$, $\gamma = 0.3$ and $\delta = 1$. These results indicate a relative stability of the distribution of SPG values, proportionally to the volume of the hierarchies, for a same density of graphs.

The results presented in figure 4 have been calculated in the following context: (1) a hierarchy O based on a lattice with a density of 0.66 described by (N=13000, H=7, W= 240), and (2) $\alpha = 0.3$, $\beta = 0.3$, $\gamma = 0.3$ and $\delta \in [0, 10]$. These results clearly show the relevance of our emotional parameter: in a negative mental state, the distributional analysis focuses on strong values of SPG and in a positive mental state, the distributional analysis is more uniform.

3.2   Application in Areas "Hygiene, Safety and Environment"

Our approach is currently evaluated in the context of a project[6] dedicated to Legal Intelligence within regulatory documents related to the domain "Hygiene, Safety and Environment". A first ontology of this domain has been defined[7]. In its current version, it is composed of 3776 concepts (depth = 11 ; width = 1300). The calculation of our gradient has been applied on a specific corpus which includes 1100 texts. This process indicates that (1) 30.2% of the SPG values are non-null, (2) 3.34% of the SPG values are equal to 1, (3) 6.18% of the SPG values belong to [0.5, 1[and (4) 63.23% of SPG values belong to]0, 0.01[. The median value of the GPS is equal to 0.128.

---

[6] This ongoing research project is funded by the French company Tennaxia (*http://www.tennaxia.com*). This "IT Services and Software Engineering" company provides industry-leading software and implementation services dedicated to Legal Intelligence.

[7] INPI June 13, 2008, Number 322.408 – SCAM-Velasquez September 16, 2008, Number 2008090075. All rights reserved.

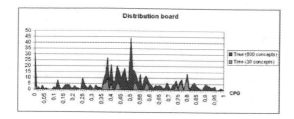

Fig. 3. Influence of the number of concepts in a tree

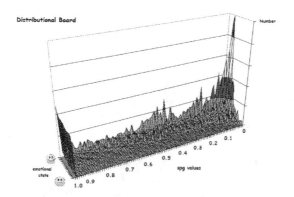

Fig. 4. Emotional parameter influence

3.3 TooPrag: A Tool Dedicated to the Pragmatics of Ontology

TooPrag (*A Tool dedicated to the Pragmatics of Ontology*) is a tool dedicated to the automatic calculation of our gradients. It takes as inputs (1) an ontology represented in OWL 1.0, where each concept is associated to a set of terms defined via the primitive *rdfs:label* and (2) a corpus composed of text files. The corpus is first indexed. Then, the ontology is loaded in memory and the SPG values of all the *is-a* links of the concepts hierarchies are computed. The LPG values of all the terms used to denote the concepts are also computed. These results are stored in a new OWL file which extends the current specification of OWL 1.0. Indeed, a LPG value is represented by a new attribute *xml:lpg* which is directly associated to the primitive *rdfs:label*. In a similar way, a SPG is represented by a new attribute *xml:spg* which is directly associated to the primitive *rdfs:subClassOf*.

4   Conclusions and Future Work

The purpose of our work, which is focused on the notion of "Personalised Vernacular Domain Ontology", is to deal with *subjectivity knowledge* via (1) its specificity to an endogroupe and a domain, (2) its ecological aspect and (3) the prominence of its emotional context. This objective leads us to study the pragmatic dimension of an ontology. Inspired by works in Cognitive Psychology, we have defined a measure dedicated to the conceptual prototypicality which evaluates the representativeness of a concept within a decomposition. This gradient reflects the pragmatics of an ontology for knowledge (re)-using. It can be of effective help in different activities, such as:

- *Ontology Evaluation.* The SPG is a relevant indicator for judging *a fortiori* the quality of a categorisation, and consequently of a domain ontology (represented for instance in OWL). Indeed, to know which are the less typical concepts of a hierarchy (according to a context of use described by an endogroup and its universe) is a good way to wonder if these concepts are at the right place? Do we have to keep them for a given mental state? Conversely, when a concept is considered as being the most typical of a category, is it really in conformity with the judgement of the experts? And is this judgement (which is based on an *a priori* decision) the good one? In this context, what we claim is that our gradients are efficient and relevant measures in the sense that they tend to reflect the real appropriation of an ontology by an endogroup. The experts are free to confirm and to objectivize (or not) these results, in the context of a "reverse ontology engineering" process [5]. Of course, when the ontology has been developed from texts, the extensional component is very strong, and it can be interesting to equilibrate the points of view by adaptating the parameters of our gradients.
- *Information Retrieval.* The SPG can be used to classify the results of a query, and more particularly an *extended* query, according to a relevance criteria which consists in considering the most representative element of a given concept (resp. a given term) as being the most relevant result of a query expressed by a (set of) term(s) denoting this concept (resp. corresponding to this term). This approach permits a classification of the extended results from a qualitative point of view. Moreover, our approach also allows us to proportion the number of results according to the value of the gradients (*i.e.* a quantitative point of view). Thus, information retrieval becomes customizable, because it is possible to adapt the results to the pragmatics of the ontology, *i.e.* privileging the intensional dimension (and not the extensional one) or conversely, working with different mental states, etc. In this way, *Ontology Personalisation* is used as a means for Web - and Semantic Web - Personalisation.

References

1. Yeung, C.M.A., Leung, H.F.: Formalizing typicality of objects and context-sensitivity in ontologies. In: AAMAS 2006: Proceedings of the fifth international joint conference on Autonomous agents and multiagent systems, pp. 946–948. ACM, New York (2006)
2. Yeung, C.M.A., Leung, H.F.: Ontology with likeliness and typicality of objects in concepts. In: Embley, D.W., Olivé, A., Ram, S. (eds.) ER 2006. LNCS, vol. 4215, pp. 98–111. Springer, Heidelberg (2006)

3. Bluck, S., Li, K.: Predicting memory completeness and accuracy: Emotion and exposure in repeated autobiographical recall. Applied Cognitive Psychology (15), 145–158 (2001)
4. Gabora, D.L.M., Eleanor Rosch, D., Diederik Aerts, D.: Toward an ecological theory of concepts. Ecological Psychology 20(1-2), 84–116 (2008)
5. Gomez-Perez, A., Fernandez-Lopez, M., Corcho, O.: Ontological Engineering. In: Advanced Information and Knowledge Processing. Springer, Heidelberg (2003)
6. Gruber, T.: Toward principles for the design of ontologies used for knowledge sharing. In: Guarino, N., Poli, R. (eds.) Formal Ontology in Conceptual Analysis and Knowledge Representation, Deventer, The Netherlands, Kluwer Academic Publishers, Dordrecht (1993)
7. Harnad, S.: Categorical perception. Encyclopedia of Cognitive Science LXVII(4) (2003)
8. McEvoy, M.E., Nelson, D.L.: Category norms and instance norms for 106 categories of various sizes. American Journal of Psychology 95, 462–472 (1982)
9. Mikulinger, M., Kedem, P., Paz, D.: Anxiety and categorization-1, the structure and boundaries of mental categories. Personnality and individual differences 11(11), 805–814 (1990)
10. Ogden, C.K., Richards, L.A.: The Meaning of Meaning: A Study of the Influence of Language Upon Thought and of the Science of Symbolism, Harcourt (1989), ISBN-13: 978-0156584463
11. Park, J., Nanaji, M.: Mood and heuristics: The influence of happy and sad states on sensitivity and bias in stereotyping. Journal of Personality and Social Psychology (78), 1005–1023 (2000)
12. Peirce, C.S.: The Essential Peirce: Selected Philosophical Writings, pp. 1893–1913. Indiana University Press (1998) (paperback)
13. Resnik, P.: Using information content to evaluate semantic similarity in a taxonomy. In: 14th International Joint Conference on Artificial Intelligence (IJCAI 1995), Montral, August 1995, vol. 1, pp. 448–453 (1995)
14. Rosch, E.: Cognitive reference points. Cognitive Psychology (7), 532–547 (1975)
15. Tversky, A., Kahneman, D.: Judgment under uncertainty: Heuristics and biases. Science (185), 1124–1131 (1974)

Ontology Evaluation through Text Classification

Yael Netzer, David Gabay, Meni Adler, Yoav Goldberg, and Michael Elhadad

Department of Computer Science
Ben Gurion University of the Negev
POB 653 Be'er Sheva, 84105, Israel
{yaeln,gabayd,adlerm,yoavg,elhadad}@cs.bgu.ac.il

Abstract. We present a new method to evaluate a *search ontology*, which relies on mapping ontology instances to textual documents. On the basis of this mapping, we evaluate the adequacy of ontology relations by measuring their classification potential over the textual documents. This data-driven method provides concrete feedback to ontology maintainers and a quantitative estimation of the functional adequacy of the ontology relations towards search experience improvement. We specifically evaluate whether an ontology relation can help a semantic search engine support exploratory search.

We test this ontology evaluation method on an ontology in the Movies domain, that has been acquired semi-automatically from the integration of multiple semi-structured and textual data sources (*e.g.*, IMDb and Wikipedia). We automatically construct a domain corpus from a set of movie instances by crawling the Web for movie reviews (both professional and user reviews). The 1-1 relation between textual documents (reviews) and movie instances in the ontology enables us to translate ontology relations into text classes. We verify that the text classifiers induced by key ontology relations (genre, keywords, actors) achieve high performance and exploit the properties of the learned text classifiers to provide concrete feedback on the ontology.

The proposed ontology evaluation method is general and relies on the possibility to automatically align textual documents to ontology instances.

1 Introduction

In this work, we present a new method to evaluate a *search ontology* [1]. The ontology supports a semantic search engine, which enables users to search for movies and songs recommendations in the entertainment domain. Semantic search corresponds to a shift in Information Retrieval (IR) from focus on navigational queries and document ranking to the higher level goals of content extraction, user goal recognition and content aggregation [2][3].

Our search engine operates in a limited domain (entertainment, movies). It relies on an explicit internal ontology of the domain, which captures a structured representation of objects (movies, actors, directors, etc). The ontology is aquired and maintained semi-automatically from semi-structured resources (such

L. Chen et al. (Eds.): APWeb and WAIM 2009, LNCS 5731, pp. 210–221, 2009.

as IMDb and Wikipedia). The ontology supports improved search experience at different stages: content indexing, query interpretation, search result ranking and presentation (faceted search, aggregated search result presentation and search result summarization).

We focus in this paper specifically on evaluating the quality of the ontology as it impacts the search process. As noted by [4], one can distinguish ontology evaluation methods at three levels: structural (measure properties of the ontology viewed as a formal graph), usability (how is the ontology accessed - through API or search tools, versioned, annotated and licensed) and functional (which services does the ontology deliver to applications). The method we present addresses *functional evaluation,* that is, we investigate how one can measure the adequacy of an ontology to support a semantic search engine.

As part of this functional evaluation, we distinguish two forms of information needs expressed by users: fact finding (the user expects to retrieve a precise set of results or to navigate to a specific movie), and exploratory search (the user seeks recommendations for several movies according to non-specific requirements). The ontology provides services to the application for both types of information needs, but in this paper, we focus on support for exploratory search.

The key idea of our evaluation method is that one can evaluate the functional adequacy of an ontology by investigating a corpus of textual documents anchored to the ontology. The textual documents are collected automatically and associated to ontology instances. Hypotheses about the ontology can then be transformed into classification tests on the corpus.

The rest of the paper is organized as follows: we first review previous work in ontology evaluation and ontology-based information retrieval (ObIR). We then present our ontology evaluation method and a set of experiments we ran to evaluate the functional adequacy of our ontology in the entertainment domain. The experiments validate the adequacy of the specific ontology acquired as part of our semantic engine for exploratory search, and provide specific, concrete indications on how to improve the ontology.

2   Dimensions of Ontology Evaluation

Evaluation of ontologies is designed and performed according to two main scenarios: assessing the quality of an ontology (by its developers) and ranking ontologies in order to choose the most suitable one for a particular task.

As a general task, evaluation of ontologies is complicated, since ontologies vary in their domain, size, purpose, language and more. Therefore, it is not possible to define a general ontology evaluation paradigm. In addition, the ontology evaluation process depends on the way the ontology was constructed: ontologies may be hand-carved, constructed by scholars or domain experts, or may be the product of an automatic or semi-automatic process. In that case, ontology quality is best measured in terms of cost/profit effectiveness.

Ontology evaluation can focus on one or more of the following dimensions:

- *Functionality (task-based)*: measures how well an ontology serves its purpose as part of a larger application;
- *Usability based*: assesses the pragmatic aspects of the ontology, *i.e.*, metadata and annotation [5];
- *Structural evaluation*: identifies structural properties of the ontology viewed as a graph-like artefact [6].

Among evaluation methods, we distinguish *extrinsic* and *intrinsic* methods. Extrinsic evaluation requires either external information in order to evaluate qualities of the ontology, such as a corpus that represents the domain knowledge (data-driven evaluation), expert opinion, or it requires a particular task which defines the context of the evaluation. Intrinsic evaluation reflects the quality of the ontology as a standalone body of knowledge. Naturally, intrinsic evaluation reflects mostly the structural properties of the ontology.

3   Search Ontologies

The usage of an ontology in our current project is motivated by the wish to improve the search experience, *i.e.*, we are interested in evaluating a search ontology as defined in the scope of ObIR (Ontology-based Information Retrieval).

The notion of semantic search refers to search techniques which go beyond the mere appearance of query words in possibly relevant documents, and aims to capture a deeper representation of the searched space and the knowledge embedded in it. Although search is widely used in the Internet, user satisfaction studies indicate that about half of the users complain about irrelevant search results (low precision) or complain about obtaining too many results (see for instance [7]). The usage of an ontology will better address user's expectations, however, it is restricts the scope of a search engine to a specific domain. In our case, we investigate the entertainment domain. For such limited scope search, semantic technology will help the engine find more relevant documents by using links among concepts (*e.g.*, movies with the same actor, similar plot), cluster results along semantic attributes to improve navigation (faceted search), and for conceptual indexing (search for "spy" and get "james bond") [3].

In order to refine the definition of evaluation of a search ontology, we refer first to distinct types of search, which represent different types of information needs (following [8][9]):

- Fact finding: a precise set of results is requested. The amount of retrieved documents is not important (for instance, a specific movie in the entertainment domain). This may correspond to a return visit to a site or a short search session.
- Exploration: the user's need is to obtain a general understanding of the search topic: high precision or recall is not required. For instance, the user explores a movies repository to find interesting movies according to his current mood or similarity with known movies.

- Comprehensive search: the task is to find as many documents as possible on a given topic (high precision and recall), and to organize the resulting set in a synthetic manner. This task is also called "briefing".

According to these information need distinctions, [9] propose a set of evaluation measures for a search ontology:

- Generic quality evaluation: checks that the ontology is syntactically correct and that it is closely related to the domain.
- Search task fitness: a different measure is applied for each search task. Measures are taken with respect to a cluster of concepts. Fact-finding fitness for a cluster of concepts is a function of the number of instances, properties and data types of all concepts in the cluster. Exploratory search fitness is a function of the number of subclasses, and Comprehensive search fitness is a function of the number of object properties, sub- and super-classes and siblings. (In all cases, the numbers are divided by the number of concepts in the cluster).
- Search enhancement capability measures how useful the ontology is for query expansions, which improve recall and precision. Recall enhancement capability is a function of the number of labels, equivalent classes, intersections and unions of concepts in a given cluster. Precision enhancement capability is a function of the number of all OWL set operations, and of the number data and object properties of concepts in a given cluster.

Such metrics are useful to evaluate ontologies in the same sense that code complexity metrics are useful when developing software. They correspond to what we call intrinsic measures above. These metrics capture the intuition that the search ontology properly supports the operation of a search engine. But these measures do not provide concrete feedback on the functional adequacy of the ontology to the domain. To illustrate the limitations of such intrinsic measures, it is possible to design an ontology to obtain high scores on all metrics with no knowledge of the domain, in a completely artificial manner, by optimizing the distribution of ontology instances across classes. To reuse the software development analogy, code complexity measures are useful to identify "bad code" (functions that are too long for example), but they do not help to assess the correctness or robustness of the code.

Beyond such metrics, we wish to define functional quality criteria for search ontologies. [3] defines the following desirable properties in a search ontology:

- Concept familiarity: the terminology introduced by the ontology is strongly connected to users terms in search queries.
- Document discrimination: the concept granularity in the ontology is compatible with the granularity used in users' queries. This granularity compatibility allows good grouping of the search results according to the ontology concept hierarchy.
- Query formulation: the depth of the hierarchy in the ontology and the complexity and length of user queries should be compatible.

– Domain volatility: the ontology should be robust in the presence of frequent updates.

This classification of functional quality criteria is conceptually useful, but it does not provide a methodology or concrete tools to evaluate a given ontology. This is the task we address in this paper.

The evaluation methodology we introduce relies on the fact that given an ontology instance (in our domain, a movie), we can automatically retrieve large quantities of textual documents (movie reviews) associated to the instance. On the basis of this automatically acquired textual corpus, we can perform automatic linguistic analysis that determines whether the ontology reflects the information we mine in the texts.

Note that we focus on evaluating the ontology itself and its adequacy to the domain as a search ontology. However, we do not simulate the search process or measure specifically how the ontology affects steps in search operation (such as indexing, query expansion, result set clustering). Accordingly, the evaluation we suggest, although informed by the task (*i.e.*, we specifically evaluate a search ontology), is not a task-based evaluation.

4   Experimental Settings: An Ontology for Semantic Search in the Entertainment Domain

We illustrate our ontology evaluation method in the context of the entertainment domain. We first describe quantitative on the experiments we have run. Our project involves the semi-automatic acquisition of an ontology in the movies domain from semi-structured data sources (IMDb, Wikipedia and other similar sources). The objective of our project is to support exploratory search over a set of documents describing movies, actors and related information in the domain.

We first report on **intrinsic evaluation** metrics over the ontology we have been assessing: number of instances, relations, density. Such measures are domain-independent. Interpretation of these measures is eventually task-oriented: we compare the metrics with those established on "high-quality ontologies" in other domains. We use for this purpose the paradigm of OntoQA [10]. Following the definition of a *search ontology*, the ontology is not expected to have a deep hierarchical structure and complex (dense) relations. The basic metrics are illustrated in Table 1. Additional metrics (instance density, relation density) confirm the expectation that the search ontology we assess has a wide and shallow structure.

Extrinsic evaluation. considers the two main search types we identified as our target scenario: fact finding and exploratory. In the first scenario, fact-finding search, the user seeks precise results and knows what she should get, the main services expected from the ontology are:

– Produce high precision results and wide coverage for terms used in the queries.
– Provide Named entity recognition functionality to allow fuzzy string matching and identify terminological variations.

Table 1. Basic measures of Ontology

Classes	33
Class instances	351,066
Relations	27
Relation instances	19
Movies	8,446
Persons	116,770

- Identify anchors, *i.e.*, minimal facts that identify a movie (for example, its title, publication year, main actors, main keywords).

For the second scenario, exploratory search, precision and recall cannot be measured since the user does not know apriori what he expects to get. Different criteria have been proposed to assess the quality of an exploratory search system [11]. As mentioned above, we do not attempt a full task-based evaluation, and, therefore, exact quality criteria for exploratory search we identify specific ways through which the ontology can improve the user experience. The services expected from the ontology are:

- Cluster instances by similarity
- Present result-sets using a faceted search GUI to provide efficient browsing and query refinement
- Identify paths of exploration through which movies are identified (period, genre, actors,)

Our task is to assess the adequacy of a specific ontology to provide the services listed above. To address this task, we adopt a corpus-based method: assume we have a corpus of textual documents associated to ontology instances. For example, for each movie instance in our ontology, we have a collection of texts. Our evaluation method translates tests on the ontology into tests on such an aligned textual corpus. We present next two specific tests illustrating this approach – to assess the ontology coverage and its classification adequacy.

5   Corpus-Anchored Ontology Evaluation

The first step of our method is to construct a corpus of documents aligned with the ontology instances. In our domain, we construct such a corpus automatically by mining movie reviews from the Web. We collected both professional, edited reviews taken from Robert Ebert's Web site[1] and additional professional and users reviews published in the Metacritic Web site[2] and 13 similar Web sources. The key metadata we collect for each document is a unique identifier indicating to which movie the text is associated. The corpus we constructed for these experiments contains 11,706 reviews (of 3,146 movies). It contains 8.7M words, with an average of 749 words per review.

---

[1] http://rogerebert.suntimes.com
[2] http://www.metacritic.com

5.1   Assessing the Ontology Coverage

To assess the fitness of our ontology to support fact-finding search, we measured the named-entity coverage of the ontology, using the constructed text corpus as reference.

We first gathered a collection of potential named-entity labels in the corpus. In professional reviews, named entities are generally marked in the *html* source. Users' reviews are not edited nor formatted. For such reviews, we relied on Thomson Reuters' OpenCalais[3] named entity recognizer to tag named entities in the corpus.

We then extracted all person names from the textual corpus and searched the labels for each entity in the ontology.

Results show that 74% of the named-entity that appear in professional reviews appear in our ontology. For user reviews (non-edited), the figure is 50%.

The main reasons for mismatches lay in orthography variations (such as accents or transliteration differences), mention of people not related to movie and aliasing or spelling variations (mostly in users reviews). We conclude that the coverage of people's names in ontology is satisfactory; however this test did not take into account variations in names and spelling that are expected.

To investigate terminological variation, we measured the ambiguity level of named-entity labels. By ambiguity, we refer to the possibility that a single name refers to more than one ontology instance. We also measured the level of terminological variation for each ontology instance – that is, given a single ontology instance (*e.g.*, an actor), how many variations of its name are found in the corpus. To identify variations in the text, we used the StringMetrics similarity matching library (http://www.dcs.shef.ac.uk/šam/stringmetrics.html). We experimented with the Levenstein, Jaro-Winkler and q-gram similarity measures. For example, using such similarity measures, we could match "Bill Jackson" with "William Jackson".

We have tested coverage on a version of the ontology that included 117,556 instances referring to persons. While taking into account only surnames, we found that 83% of the names are ambiguous. There are 18.57 variations on average for each ontology instance.

This simple exercise indicates how a textual corpus aligned with the ontology and mature language technology (named-entity recognition and flexible string similarity methods) allows us to measure a complex property of the ontology. This evaluation does not only provide a score for the ontology. It also indicates which specific named entities are used in the corpus, how often, which confusions can be expected when disambiguating query terms and how to specifically improve the terminology-related services provided by the ontology.

In the next section, we demonstrate how the more complex task of measuring the clustering adequacy of the ontology can also be assessed using text classification techniques.

---

[3] http://www.opencalais.com

5.2   Assessing the Classification Fitness of an Ontology

As discussed above, the fitness of the ontology to support exploratory search is a function of the number of subclasses. We take this definition a step forward: the number of subclasses is valid if it produces a balanced view of the world domain (represented by the documents) and if the explicit characteristics of the hierarchy can be identified implicitly in the documents.

An ontology induces a hierarchical classification over its elements. Each class (*e.g.*, actor, genre) may be viewed as a dimension for classification of the texts that represent the domain. The ontology provides effective classification services if it meets two criteria:

- The Ontology classification is **useful** if the induced classification is well-balanced, enabling the user explore the dataset in an efficient manner (for exploratory purposes).
- The Ontology classification is **adequate** if the classification induced by the ontology is valid with respect to the domain, which is represented by texts.

Accordingly, we formulate the following hypothesis:

Hypothesis. *If* the ontology indicates that some movies are "clustered" according to one of the dimensions, *then* documents associated to these movies should also be found to be associated by a text-classification engine that has been trained on the classification induced by the ontology.

The general procedure we performed to test this hypothesis is the following:

Step 1: Choose a dimension to test (we have tested genre, actors and keywords).
Step 2: Induce a set of categories (subsets of movies). The subclasses of this dimension and the films instantiated under each subclass defines a clustering of the movies. For example, if we evaluate the "genre" dimension, we cluster movies according to their genre property. In our ontology, this produces about 30 classes of movies (one for each genre value).
Step 3: Gather texts (from the reviews corpus, texts that were not used in the acquisition process of the ontology) related to these movies and form a collection (Text_{ij}, movie_i).
Step 4: Train a classifier on a subset of the texts (Text_{ik}, movie_i, category_i) where category_i is the category induced by the ontology.
Step 5: Test the trained classifier on withheld data (Text_{ij}, movie_i) and compute accuracy, precision and recall with respect to the category.

Hypothesis. Adequate classes yield high accuracy and F-measure on an instance-aligned corpus.

5.3   Parameters

There are several reasonable options to perform the text classification task in Step 4 above, with different methods of text representation and with different classifiers.

For text representation, we viewed texts as "bag of words", *i.e.*, as unigrams, and represented each text as a Boolean vector in which each coordinate indicates the existence, or lack of existence, of a string in the text. We tested a few options of pre-processing on the texts and of selecting the features (the strings that we take into account when representing the text): with and without stemming[4] and with and without filtering noise words; selecting features using Mutual Information (MI), or using TF/IDF; and with different numbers of features top 300 or 1000.

Mutual Information-based feature selection is inspired by [12] which shows that this method yields best results on text categorization by topic on a standard News corpus.

The feature selection methods we used are as follows: in TF/IDF, words with the highest values were chosen as features, for the entire corpus. In MI, the features with the highest mutual information associated with the class were chosen (a different set of features is used for every class).

For the classifying task, we used two methods: Support Vector Machines (SVM) (linear and quadratic) and Multinomial Naïve Bayes (MNB) as implemented in the Weka toolkit [13].

5.4   Results

We applied the classification procedure to the classification induced by the genre dimension. The classifiers were trained on the reviews corpus. We performed 5-fold cross-validation on the corpus.

The best text representation was established by testing the genre classifier on the task of classification of one class against all.

16 different experimental settings were tested:

- TF/IDF vs. MI.
- Vectors of size 300 vs. 1000 features.
- Stemmed words vs. Raw.
- Noise words filtered vs. no filtering.

For each possibility we tested both SVM and Naïve Bayes as classifiers.

Classification by Genre. Genres, according to IMDb.com are defined to be "simply a categorization of certain types of art based upon their style, form, or content. Most movies can easily be described with certain umbrella terms, such as Westerns, dramas, or comedies". The tested ontology includes 23 genre subclasses.

We performed the classification process as described above, and found that the best combination is MI, 1000 features, no stemming, noise filtering, and Naïve Bayes as classifier. The Average F-Measure is 0.41 (all results shown in Table 2). It is possible to explain the failure of the SVM to outperform the Naïve Bayes classifier, due to the imbalanced size of the classes, as shown in [14].

---

[4] We used the classical Porter Stemmer for the experiment.

Table 2. F-Measure of classification engine *One-vs-All*

Genre	F-Measure
Drama	0.841
Sport	0.719
Comedy	0.709
Thriller	0.682
Family	0.626
Adventure	0.625
Action	0.616
Documentary	0.613
Sci-Fi	0.551
Horror	0.540
Animation	0.539
Fantasy	0.533
Music	0.500
Crime	0.490
Romance	0.462
Western	0.431
Mystery	0.352
History	0.289
Musical	0.274
War	0.257
Short	0.239
Biography	0.231
Adult	0.198

Table 3. Pair Classification of Genres

Pair	F-Measure	Accuracy
Drama - Western	0.997	0.994
Drama - Musical	0.996	0.991
Thriller - Musical	0.986	0.972
Action - Western	0.979	0.960
Thriller - Western	0.977	0.956
Action - War	0.935	0.894
War - Action	0.438	0.809
Adult - Romance	0.367	0.773
Biography - Documentary	0.358	0.739
History - Short	0.343	0.821
Adult - Short	0.287	0.702
Biography - Drama	0.172	0.903

The results indicate that some genres are very well defined (drama, sport, comedy), while others cannot be recovered by analyzing the text of the reviews (musical, short, biography, adult).[5] While these figures provide a first assessment of the quality of each genre category, pair-wise classification provides finer-grained tests of the level to which pairs of genres can be distinguished. A subset of the results showing best and worst cases is shown in Table 3. We report both F-measure and Accuracy for these tests.

The average error rate for pairwise classification is 16.2%; it varies significantly between genre pairs, and therefore can indicate a weak category or classes which are harder to differentiate.

---

[5] Specifically, the genres of music and musical are derived from the IMDb genres and are apparently confusing.

For comparison, we have tested a Baseline classifier which is not related to the Ontology under test in any way. This was done by creating 25 random classes of 1,000 movies. We performed the same classification procedure. The results showed average F-measure lower than 0.16 (as opposed to 0.41 overall for the ontology-based classifiers, and over 0.70 when we filter out low-quality genres) and extremely low accuracy (less than 0.1). This indicates that the corpus-anchored ontology evaluation method does not capture random patterns of text classification.

Note that the pair-wise classifiers are not symmetric: this is because there can be overlap between two categories. For example, a movie can belong both to the genres of action and drama. In our experiment, when we test the pair drama-action, we learn a binary classifier that responds "true" for texts classified as drama, and "no" for all the rest. This classifier is only trained over documents associated to movies that are tagged as either drama or action (all other texts are ignored). If a movie is tagged as both drama and action, it will be classified as "true" for the drama-action classifier as well as for the action-drama classifier. This asymmetry provides an indication that one genre may be included in another.

6   Conclusion and Future Work

We have presented a concrete ontology evaluation method based on the usage of a corpus of textual documents aligned with ontology instances. We have demonstrated how to operate such evaluation in the case of an ontology in the entertainment domain used to improve a semantic search engine.

We have first constructed an ontology-aligned textual corpus by developing a Web crawler of movie reviews.

Our first experiment measures the adequacy of the ontology to support fact-finding search. We have found specifically that our ontology has wide coverage but lacks support for ambiguity resolution and terminological variation handling. We use human-language technology to translate hypothesis on the ontology coverage into measures of properties of the textual.

Our second experiment measures the adequacy of the ontology to support exploratory search. We have formulated hypotheses that capture the quality criteria of an exploratory search system, and tested these hypotheses on our ontology-aligned textual corpus. Specifically when testing the classification adequacy of our ontology along the "genre" dimension, we found that most of the genres in the ontology induce high-quality text classifiers - but some, such as sport and music) do not induce appropriate classifiers. This method provides specific feedback to the ontology maintainer.

Our tests support the claim that classification as a method for evaluation is adequate.

Acknowledgments

This research was supported by Deutsche Telekom Laboratories at Ben-Gurion University of the Negev. We wish to thank Jon Atle Gulla and Jin Liu for fruitful collaboration.

References

1. Burkhardt, F., Gulla, J.A., Liu, J., Weiss, C., Zhou, J.: Semi automatic ontology engineering in business applications. In: Proceedings of the 3rd International AST Workshop – Applications of Semantic Technologies. LNI, vol. 134, pp. 688–693 (2008)
2. Baeza-Yates, R., Ciaramita, M., Mika, P., Zaragoza, H.: Towards semantic search. In: Kapetanios, E., Sugumaran, V., Spiliopoulou, M. (eds.) NLDB 2008. LNCS, vol. 5039, pp. 4–11. Springer, Heidelberg (2008)
3. Gulla, J.A., Borch, H.O., Ingvaldsen, J.E.: Ontology learning for search applications. In: Meersman, R., Tari, Z. (eds.) OTM 2007, Part I. LNCS, vol. 4803, pp. 1050–1062. Springer, Heidelberg (2007)
4. Gangemi, A., Catenacci, C., Ciaramita, M., Lehmann, J.: Modelling ontology evaluation and validation. In: Sure, Y., Domingue, J. (eds.) ESWC 2006. LNCS, vol. 4011, pp. 140–154. Springer, Heidelberg (2006)
5. Gomez-Perez, A.: Evaluation of ontologies. International Journal of Intelligent Systems 16, 391–409 (2001)
6. Alani, H., Brewster, C.: Ontology ranking based on the analysis of concept sructures. In: Proceedings of the 3rd International Conference on Knowledge Capture (K-Cap), Banff, Canada, pp. 51–58 (2005)
7. JupiterResearch: Search technology buyers guide. Technical report, IBM Content Discovery (2006),
ftp://ftp.software.ibm.com/software/data/cmgr/pdf/searchbuyersguide.pdf
8. Aula, A.: Query formulation in web information search. In: Proceedings of IADIS International Conference WWW/Internet, pp. 403–410 (2003)
9. Strasunskas, D., Tomassen, S.L.: Empirical insights on a value of ontology quality in ontology-driven web search. In: Meersman, R., Tari, Z. (eds.) OTM 2008, Part II. LNCS, vol. 5332, pp. 1319–1337. Springer, Heidelberg (2008)
10. Tartir, S., Arpinar, I., Moore, M., Sheth, A., Aleman-Meza, B.: OntoQA: Metric-based ontology quality analysis. In: Proceedings of Workshop on Knowledge Acquisition, Autonomous, Semantically Heterogeneous Data and Knowledge Sources, pp. 45–53 (2005)
11. White, R.W., Muresan, G., Marchionini, G. (eds.): ACM SIGIR Workshop on Evaluating Exploratory Search Systems, Seattle (2006)
12. Dumais, S., Platt, J., Heckerman, D., Sahami, M.: Inductive learning algorithms and representations for text categorization. In: Proceedings of the Seventh international Conference on Information and Knowledge Management, Bethesda, Maryland, pp. 2–7 (1998)
13. Witten, I.H., Frank, E.: Data Mining: Practical machine learning tools and techniques, 2nd edn. Morgan Kaufmann, San Francisco (2005)
14. Akbani, R., Kwek, S.S., Japkowicz, N.: Applying support vector machines to imbalanced datasets. In: Boulicaut, J.-F., Esposito, F., Giannotti, F., Pedreschi, D. (eds.) ECML 2004. LNCS (LNAI), vol. 3201, pp. 39–50. Springer, Heidelberg (2004)

Core-Tag Clustering for Web 2.0 Based on Multi-similarity Measurements*

Yexi Jiang, Changjie Tang, Kaikuo Xu, Lei Duan, Liang Tang,
Jie Gong, and Chuan Li

School of Computer Science Sichuan University,
Chengdu, 610065 China
{jiangyexi,tangchangjie}@cs.scu.edu.cn

Abstract. Along with the development of Web2.0, folksonomy has become a hot topic related to data mining, information retrieval and social network. The tag semantic is the key for deep understanding the correlation of objects in folksonomy. This paper proposes two methods to cluster tags for core-tag by fusing multi-similarity measurements. The contributions of this paper include: (1) Proposing the concept of core-tag and the model of core-tag clusters. (2) Designing a core-tag clustering algorithm CETClustering, based on clustering ensemble method. (3) Designing a second kind of core-tag clustering algorithm named SkyTagClustering, based on skyline operator. (4) Comparing the two algorithms with modified K-means. Experiments show that the two algorithms are better than modified K-means with 20-30% on efficiency and 20% higher scores on quality.

Keyword: folksonomy, tag, clustering, clustering ensemble, skyline.

1 Introduction

With development of Web information technology, the available resources have increased dramatically. *Taxonomy* is used to classify the online resources. However, in most cases of Taxonomy, resources are categorized by experts, thus they cannot reflect the original opinions of the users. To solve these problems, the authors of [1, 2] proposed new concept named *folksonomy*. It allows users to add metadata in the form of keywords to shared resources. Folksonomy [3] allows users effectively organize and share vast amount of information. Intuitively, folksonomy is a tripartite graph, in which users, tags and resources are nodes while relations among them are edges. It reflects users' true opinions on resources via collaborative intelligence. The core of folksonomy is defining tags by users. Since it would cause a large number of different tags for similar resources, an efficient clustering mechanism to put tags together is necessary to get related tags by giving query tag. We call the query tag **core-tag**. In this paper, we focus on handling core-tag clustering. Example 1 illustrates the idea of core-tag clustering.

---

* Supported by the 11th Five Years Key Programs for Sci. & Tech. Development of China under grant No. 2006BAI05A01, the National Science Foundation under grant No. 60773169, the Software Innovation Project of Sichuan Youth under grant No. 2007AA0155.

L. Chen et al. (Eds.): APWeb and WAIM 2009, LNCS 5731, pp. 222–233, 2009.

Example 1. A user wants to find out the related tags of 'web2.0' and tries to know their closeness to the core tag but without trivial details. Suppose the related tag set be {blog, Social, library2.0, Design, Web, community, collaboration, mashup, online, webdesign, socialnetworking, video, education, ajax, aggregator, search techonology, Blogs, wiki, rss}. The clustering result can be seen in Table 1 and Fig. 1.

Table 1. Result of Core-tag Clustering

Clustering result by using core-tag 'web2.0'			
Group 1:	blog, Social, library2.0;	Group 2:	Design, Web, community;
Group 3:	collaboration, mashup;	Group 4:	online, webdesign, video;
Group 5:	education, aggregator;	Group 6:	search, technology, Blogs, wiki;

Clearly, the tags in a group are related with each other. Moreover, we can see that the group with smaller index is closer (to the core tag) than the group with larger index. The distance between a cluster and the core tag indicates their semantic similarity.

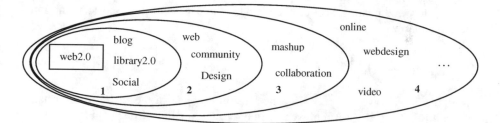

Fig. 1. Core-tag Clustering using core-tag

It is a challenging task with following difficulties:

(a) The existing association rule mining algorithms can find the correlated tags, but cannot efficiently group the correlated tags.
(b) Some existing clustering algorithms work well to find the groups, but clusters are unordered and cannot be distinguished.
(c) Some algorithms like K-means are not stable. They lead to the randomness of result. For different runs the result may be different. The randomness would lower the accuracy of the result.
(d) Some algorithms require a pre-fix parameter to set the number of clusters. It depends much on the subjective of the users. Such parameter settings are usually empirically set and difficult to be determined.

To overcome above problems, we introduce two core-tag clustering methods with different ideas. That is: (a) The idea of clustering ensemble and (b) the idea of sky-line. A formal definition of core-tag clustering is as follows:

Problem Statement. Given a core-tag T_c and set S that contains candidate tags related to T_c. Find a set of clusters contain at least one of the most related tags, let C denotes the set of clusters. The result should satisfy the following conditions:

- $\cup C \subseteq S$ and $|\cup C| \leq k$. Variable k denotes the cardinality of S.
- For $\forall C_j \in C$, there have $C_j \neq \varnothing$ and $C = C_1 \cup C_2 ... \cup C_n$. Variable C_i denotes a certain cluster.
- For $\forall C_i, C_j \in C$, there have either $\text{dist}(T_c, C_i) < \text{dist}(T_c, C_j)$ or $\text{dist}(T_c, C_i) > \text{dist}(T_c, C_j)$. The function $\text{dist}()$ denotes the distance from the core-tag to a cluster by considering the distance of all tags in the cluster.

The contributions of this paper include: (1) Proposing the concept of core-tag and the model of core-tag clustering based on multi-similarity measurements. (2) Designing core-tag clustering algorithm CETClustering based on clustering ensemble method. (3) Designing core-tag cluster algorithm SkyTagClutering based on skyline operator. (4) Analyzing the best method based on experiment results both on the efficiency and the quality of the clustering result.

The rest of the paper is organized as follows: Section 2 describes the related work. Section 3 proposes algorithm CETClustering, based on clustering ensemble method. Section 4 proposes algorithm SkyTagClustering, based on skyline operator. Section 5 designs experiments to evaluate the two methods and compares them with K-means on efficiency and quality. Section 6 concludes the paper with a discussion and introduces the future work.

2 Related Work

Ensemble method was proposed for data fusion problem in the realm of information retrieval [6]. Then it was employed to improve the quality of decision tree in Quinlan's paper [8]. This method was widely applied in classifying. Literature [9], [10] and [11] introduced how this method be used in clustering field. N.C. Oza [9] gave an example. It used a set of K-means algorithm as basic algorithms and used the ensemble method to combine the results of these basic algorithms. A.Strehl and J.Ghosh proposed a framework of a classical clustering ensemble process in [10]. The key idea is as follows: (a) Clustering the data independently by a set of clusterers. (b) Constructing a graph based on the result. (c) Decomposing the graph into proper number of parts to get final result. In [11] the authors gave an experiment to analyze the quality of the method by counting the mis-assigned objects.

Consider the skyline operator method. Top-k problem is a classical problem in database field. It first appeared in S. Borzsony's paper [12]. It proposed a complicated but not so efficient algorithm for calculating the skyline. Literature [13] introduced two progressive algorithms to compute the skyline: The Bitmap version and the Index version. The Bitmap version is efficient for computing the skyline with discrete values. The Index version used a B-tree to help store the information and sort the element in each dimension to help get the skyline. They are efficient; however, they both need some prerequisite, such as: (a) Skyline operation involves all the dimensions of the data. (b) All the values are within the range [0, 1]. (c) Each dimension has discrete value. Unfortunately, the tag similarity datasets can only fulfill the first two requirements even after the preprocessing. For each dimension in our similarity set that represents the result of a certain similarity method, the value of the similarity is continuous and has the accuracy of 4 digit after the point, thus if the algorithm in [13] is employed, there would be thousands of distinct values, which is prohibitive. In [14] and [15] the authors mainly focused on computing the skyline when elements has high dimension. For [14], it introduced the concept skyline frequency that compares and ranks the interestingness of data points based on how often they are returned in the skyline when different number of dimensions. Literature [15] introduced another new concept k-dominant. Their proposal of k-dominance offers a different notion of interestingness from skyline frequency. It focused on how many dimension an element E1 dominate E2. Authors of [16] developed BBS (Branch and Bound Skyline), a progressive algorithm based on nearest neighbor search, which is IO optimal. It employs an R-tree to help store the needed information and avoid the duplicates. The skyline algorithm in [17] can quickly return the first result and produce more and more results continuously. That means the algorithm can get part of the skyline before the whole process of computation.

3   The Algorithm CETClustering

In order to solve core-tag clustering problems, we propose an algorithm based on the membership of tags in the same cluster. [10] introduces a framework for combining multiple partitions of multiple clusterers by using the graph partitioning method. We use the same framework in our work. However, the graph partition methods they used are not suitable for the core-tag clustering, we use our own partition method by modify minimum spanning tree to maximum spanning forest. The reasons lies that: (a) The methods can only solve bi-partitioning problem. In tag clustering situation, it needs to partition the graph into any number of parts. (b) They aim to partition huge graph while in tag clustering problem the graph is much smaller.

The key steps of core-tag clustering are as follows:

Step 1: Use a number of clusterers to cluster tags independently.
Step 2: Generate the co-association matrix CoA according to the clustering result of each clusterers.
Step 3: Decompose the graph represented by the CoA to any number of parts wanted.

3.1   Preprocess

Before using CETClustering algorithm, we need to do some preprocess to the data in order to fulfill the input requirement.

In [4], we measured the similarity between the tags by eight formulas (AEMI, Simrank, etc.) Since there is no existing similarity measurement considering all the three factors altogether (user, tagged page and tag), we need to consider multiple measurements as a whole to cluster tags to improve the clustering result.

Table 2. Similarity Set

Java	Dimension 1	Dimension 2
J2EE	0.841	0.718
JVM	0.812	0.802
JDK	0.794	0.811
JRE	0.748	0
J2ME	0	0.718

The steps of preprocessing are as follows:

(1) Generate similarity set *Set* for each core-tag and remove unqualified elements. In this step, we have a set S for each core-tag T_c. It contains k n-dimensional vectors. The variable k is the cardinality of S; n is the number of similarity measurements used. Each dimension of a vector denotes the similarity value between T_c and a certain related tag calculated by a certain measurement (See Table 2). We remove those vectors with having missing value in more than half dimensions and fill remain missing values with 0. Now the similarity set *Set* is created.

(2) Normalize all the values in set *Set*. In this step we adopt min-max normalization to convert all the values into domain [0, 1].

(3) Fill in missing value in the set. The remaining missing values are filled with the average value of the other element in the vector.

	JDK	JRE	JVM	J2EE	J2ME	Eclipse	EJB
JDK	3	2	0	2	1	0	0
JRE	2	3	3	1	1	0	1
JVM	0	3	3	1	1	2	1
J2EE	2	1	1	3	2	2	3
J2ME	1	1	1	2	3	1	0
Eclipse	0	0	2	2	1	3	1
EJB	0	1	1	3	0	1	3

Fig. 2. Co-Association Matrix

3.2 Independent Clustering and Generate Co-association Matrix

In this step, we use several clusterers to cluster tags independently according to *Set*. Here we use K-means as clusterer.

Co-Association matrix CoA (Fig. 2) represents the relationship between the tags based on the results of several clusterers. First, each of the cluster result computed by clusterer k is a partition $\prod_k$, the symbol $\prod_k = \{\pi_1, \pi_2, \ldots \pi_H\}, \pi_1 \cup \pi_2 \cup \ldots \cup \pi_H = Set$. Here π_i represents a certain cluster of the result of clusterer k. Then employ Kronecker Delta Function as the consensus function to set the value of CoA. The following is the Kronecker Delta Function: Here, $\pi_i(x)$ equals to 1 means tag x is in cluster π_i, otherwise it equals to 0, otherwise, it equals to 0.

$$s(x, y) = \sum_{i=1}^{H} \delta(\pi_i(x), \pi_i(y)), \delta(a, b) = \begin{cases} 1, a \text{ and } b \text{ are both } 1 \\ 0, \text{ otherwise} \end{cases} \qquad (1)$$

The main properties of the CoA matrix are as follows:

Property 1: Each cell of CoA represents how many clusterers put the two tags into the same cluster. Its value would not exceed the number of clusterers.

Property 2: The matrix is diagonally symmsetric. That means, for all cells, CoA(i, j) always equals to CoA(j, i).

Property 3: The graph represent the matrix may not be a connected graph.

Proof. There exists possibility that these tags belong to different clusters and all the clusterers put these tags into the different clusters. (See Example 2).

Example 2. There are 3 tags: *'fish'*, *'sea'* and *'Philosophy'*. Three clusterers may put *'fish'* and *'sea'* into one cluster and put *'Philosophy'* into another, thus the CoA is not a connected. It contains two disconnected sub-graph. (Table 3 and Fig. 3)

Table 3. Cluster result of Example 2

Clusterer	Result
Clusterer 1	{*fish, sea*}, {*Philosophy*}
Clusterer 2	{*sea, fish*}, {*Philosophy*}
Clusterer 3	{*Philosophy*}, {*fish, sea*}

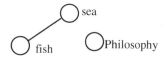

Fig. 3. Graph of Example 2

3.3 Decompose Graph

The graph can be decomposed into any number of sub-parts. It's a re-clustering based on the result of the first time clustering. The matrix represents a graph. We modify MST (Minimum Spanning Tree) algorithm to minimum spanning forest, and then decompose the forest to any number of parts according to the edge value of pairs of nodes. The pseudo code is as follows:

Algorithm 1: Decompose.

    Input: N, number of part; G, graph
1.   Let n = current partition of G and CL← Ø
2.   while n < N
3.   find the node with the smallest weight and split the edge.
4.   calculate the distance between each partition and the tag.
5.   for each partition
6.   create cluster according to partition
7.   add cluster to CL
8.   Sort CL according to distance

Note that: (a) Each node is used to record the tag itself and its parent, if it is the root of a tree, the parent is NULL, otherwise it point to another tag. (b) The graph of the matrix may not be a connected graph. (c) In Step 3–4, while the number of parts is less than expectation, the loop continually split the forest.
The whole process of the algorithm is as follows:

Algorithm 2: Clustering Ensemble.

    Input: n, clusterer size; **Set**, similarity set
1. Use n clusterers to cluster the tags according to **Set**
2. Generate CoA matrix G for $G(x, y) = \sum_{i=1}^{|C|} \delta(\pi_i(x), \pi_i(y))$
3. for each node u ∈ G
4.   do key[u] ←0 and parent[u] = NULL
5. key[r] ← 0
6. put all nodes into a priority queue Q

```
7. while Q ≠Ø
8.   do u = extract-min(Q)
9.     for each v is adjacent to u
10.      do if v ∈Q and w(u, v) > key[v]
11.        then parent[v] = u
12.          key[v] = w(u, v)
13. return Decompose(parent, N)
```

Note that: (a) In Step 1, the clusterers are used to cluster the tags. (b) CoA is created in Step 2. (c) Maximum weight among all the neighbors and the parent of node v is recorded in key[v] and parent[v] respectively through Step 3 - 12.

4 SkyTagClustering Algorithm

Algorithm 2 is good in stability, but it is still with the following deficiencies:

- The user must pre-assign a parameter to set the number of clusters. The ideal situation is to let the algorithm itself find the proper number of clusters according to the data rather than manual set beforehand.
- The quality and efficiency of the clustering results depend on the basic algorithm chosen. If the basic algorithm is not good, the performance of CETClustering would be affected.

To solve the problems, we propose new algorithm based on skyline borrowing some ideas from [15]. As the authors mentioned in [15], under relatively lower dimension (d < 12), the two-scan algorithm version (TSA) has better performance, so our method employs it as the basic algorithm. The definition of k-dominate and the pseudo code of TSA can be referred in [15]. The algorithm asks for three parameters D, S and k, respective means the dataset, in our algorithm it is *Set*, the data space and the parameter k for k-dominate.

There are several advantages for us to employ skyline as the basic algorithm:

a) Existing algorithm for skyline are relative efficient.
b) The performance of skyline is not affected by weight of dimension
c) Users do not need to control the number of clusters.

The steps of the SkyTagClustering algorithm are as follows: (1) Transform *Set* into proper form. (2) Iteratively extract skylines using basic skyline operator algorithm.

4.1 Data Transformation

We use the same preprocess method stated in Section 3 for SkyTagClustering. The similarity set *Set* should be transformed into proper form to fulfill the need of skyline operator.

In original *Set*, 1.0 means the most correlated and 0.0 means the least correlated. The transformation is easy. In new *Set*, we simply set the values as 1 minus the original value. Thus all the values denote the distance from a tag to core-tag. After transformation, the smaller the value, the closer it should be.

4.2 Iterative Extract Skylines

TSA is iteratively used to retrieve one skyline per time. Each time the skyline is found, the points in the skyline are removed from similarity set and used to create a cluster and the remaining data are reused to find new skyline until no point left (See Fig. 4).

There are some properties of the algorithm SkyTagClustering:

Property 1: The earlier found skyline is closer to the target tag.

Proof. For any tag T_n not in the skyline must be dominated by at least one tag T_s in skyline. This means there exist at least one tag in skyline more close to T_c than T_n.

Property 2: The algorithm is progressive. Thus it can quickly return the first cluster (the closest group of tags).

Proof. The skylines are iteratively computed. If only the first cluster is needed, the algorithm can return it and stop quickly.

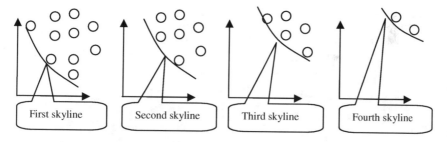

Fig. 4. Iteratively extract skylines

The pseudo code of this algorithm stated as follows:

Algorithm 3: k-dominant Skyline Clustering Algorithm(D, S, k)
---
1.   if k > D. dimension
2.   report error, k is set too large
3.   Transform the similarity set
4.   initialize set of cluster $C = \emptyset$
5.   while D != $\emptyset$
6.   cluster C_i = TSA(D, S, k)
7.   if |cluster| = 0 then report error, k is set too small
8.   remove the points in C_i from D
9.   add C_i to C
10. end while
11. return C

Note that: (a) Step 1 guarantees k > D. (b) Step 3 transforms similarity set to meet the requirement of skyline operator. (c) The skylines are iteratively computed by the help of TSA, for each skyline, a new cluster is created to put the points in. After this algorithm, a set of clusters C is formed.

5   Experiment and Performance Study

5.1   Experiment Setting

The goal of experiment is to find a good clustering algorithm for core-tag clustering. A modified K-means is applied as baseline method and we compare it with CETClustering (CE for short) and SkyTagClustering (SC for short) both on the aspects of efficiency and the quality.

Experiment Data: All the dataset are real dataset downloaded from http://del.icio.us from Nov 20 to Dec 15, 2007. There are a total of 234023 unique tags and 749971 unique web pages. All the data are processed by previous work mentioned in [4] so the processed data will be directly used.

Platform: All the experiments are implemented by java and are conducted on a PC with Intel Core2 Due CPU with 2G memory, running on Windows server 2003.

5.2   Efficiency Comparison

The efficiency of the three algorithms can be seen in Fig. 5 and 6. We set k (Size of candidate related tags) as 20 and 30 respectively, numbers of clusterers in CE as 3 (CE3) and 5 (CE5) respectively, and set the clustering tasks (number of core-tags) through 10 to 100 to test their speed. It is clear that SC has the best performance. CE

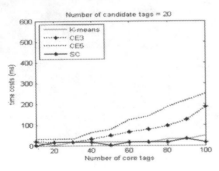

Fig. 5. Number of candidate tags = 20

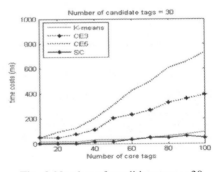

Fig. 6. Number of candidate tags = 30

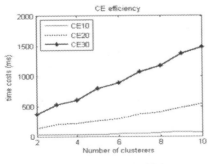

Fig. 7. CETClustering efficiency

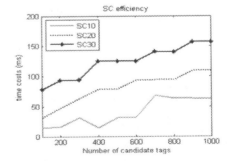

Fig. 8. SkyTagClustering efficiency

cost more time than the other two algorithms. It's quite reasonable. It costs at least five times longer than K-means version since it employs five K-means and does some extra work to get the final result. With increase number of K-means, time costs will increase undoubtedly.

Fig. 7 shows how clusterers size affects the efficiency of CE algorithm. We test the time costs by changing the number of potentially related tags for each core tag from 10 to 30 (CE10, CE20 and CE30). It is obvious that the time cost of CE is not linear increasing as the increasing of the number of clusterers, it's because besides the independent clustering of the clusterers this algorithm also does some work to combine the results of each clusterer to get final result, it would cost extra time. In efficiency experiment, CE performs worst and SC performs best.

Fig. 8 shows k affects the efficiency of SC algorithm. This algorithm is also tested by changing parameter k for each core-tag from 10 to 30 (SC10, SC20 and SC30). It can be seen that the number of candidate tags linearly affected the time costs of SC. Moreover, it is obvious that SC is a scalable algorithm, the clustering tasks increase from 100 -1000 but the time cost just doubled.

Above all, CETClustering algorithm doesn't perform well in the efficiency experiment because it employs several K-means. SkyTagClustering, on the contrary, performs very well both on the comparative and scalable test.

5.3 Remarks Clustering Result

As there is no authoritative benchmark method for research of web2.0, we ask three human to evaluate the quality of these algorithm versions: K-means, CE 3, CE 5 and SkyTagClustering. For each algorithm, there are 50 core-tags sample for evaluators to mark. For each core tag, each cluster of the result is assigned an index, the smaller the index the closer the cluster to the core tag. The evaluator should give a score according to the core tag and a set of clusters. The meaning of score is as follows:

$$S \begin{cases} 5: \text{Perfect. Tags are put into right cluster and the number of cluster is proper.} \\ 4: \text{Very good. Most of the tags are right placed.} \\ 3: \text{Good. A large part of the tags are right placed.} \\ 2: \text{Fair. Some of the tags may not be put into right cluster.} \\ 1: \text{Bad. Most of the tags are wrongly placed.} \end{cases}$$

The method to evaluate the algorithm is very simple.

$$S_a = (\sum_{i=1}^{U} \sum_{j \in a} score_{aj}) / N_a \tag{4}$$

Here, S_a means final score of algorithm a, U is the number of evaluators, $score_{aj}$ is the score for each core tag get by algorithm a, N_a is the total number of scores of a. Table 5 shows the evaluate result of the four algorithms.

Table 4 indicates that SkyTagClustering has the best quality for it owns several advantages that other algorithms lack. CETClustering with 5 K-means ranked the second and 3 K-means ranked the third for they are more robust than the single K-means

Table 4. The scores of algorithms

Algorithm	Average Scores	Standard Deviation	Rank
K-means	2.71	0.51	4
CETClustering 3	2.96	0.52	3
CETClustering 5	3.19	0.33	2
SkyTagClustering	3.48	0.61	1

version. They can reduce the randomness of single K-means, the more number of clusterer, the more stable it is. However, as the increase of the cluterers, its efficiency would be affected. As the scores of CE3 and CE5 are very near but CE3 is more efficient than CE5, combining the efficiency and quality, CE3 outgoes CE5.

6 Conclusion and Future Work

We proposed two algorithms to handle the core-tag clustering task by using multi-similarity measurements. In conclusion, SkyTagClustering is considered to be the best algorithm for it has the best performance both on efficiency (10% - 50% faster than K-means) and qualification. CETClustering has a good performance in qualification. If it is used clustering algorithm faster than K-means, it would have better speed.

The future work includes: Employ some other clustering algorithm as the basic algorithm of CETClustering to do the clustering work and evaluate its performance. Find a more efficient skyline algorithm as the basic algorithm and let the algorithm create more proper number of clusters. For the whole project, we have done the first 6 steps. Multi-tags correlation problem and tag networks are still left. Our next work is to finish these two tasks based on previous work.

References

1. Mates, A.: Folksonomies – Cooperative Classification and Communication through Shared Metadata. In: Computer Mediated Communication, LIS590CMC (2004)
2. Hammond, T., Hannay, T., Lund, B., Scott, J.: Social Bookmarking Tools:A General Review. D-Lib Magazine (2005)
3. Hotho, A., Jäschke, R., Schmitz, C., Stumme, G.: Information retrieval in folksonomies: Search and ranking. In: Sure, Y., Domingue, J. (eds.) ESWC 2006. LNCS (LNAI), vol. 4011, pp. 411–426. Springer, Heidelberg (2006)
4. Xu, K., Chen, Y., Jiang, Y., Tang, R., Liu, Y., Gong, J.: A comparative study of correlation measurements for searching similar tags. In: Tang, C., Ling, C.X., Zhou, X., Cercone, N.J., Li, X. (eds.) ADMA 2008. LNCS, vol. 5139, pp. 709–716. Springer, Heidelberg (2008)
5. Fred, A., Jain, A.K.: Evidence Accumulation Clustering based on the K-means Algorithm. In: Proceedings of the Joint IAPR International Workshop (2002)
6. Voorhees, E., Gupta, N.K., Johnson-Laird, B.: The Collection Fusion Problem. In: The Third Retrieval Conference (1995)
7. Dietterich, T.G.: Ensemble methods in machine learning. In: Kittler, J., Roli, F. (eds.) MCS 2000. LNCS, vol. 1857, p. 1. Springer, Heidelberg (2000)

8. Quinlan, J.R.: Bagging, boosting, and C4.5. In: Proc. of the13th AAAI Conference on Artificial Intelligence. AAAI Press, Menlo Park (1996)
9. Oza, N.C.: Ensemble Data Mining Methods. NASA Ame Research Center (2000)
10. Strehl, A., Ghosh, J.: Cluster Ensembles – A Knowledge Reuse Framework for Combining Partitionings. AAAI, Menlo Park (2002)
11. Topchy, A., Jain, A.K., Punch, W.: Combining Multiple Weak Clusterings. In: ICDM (2003)
12. Borzsony, S., Kossmann, D., Stocker, K.: The Skyline Operator. In: ICDE (2001)
13. Tan, K.L., Eng, P.K., Ooi, B.C.: Efficient progressive skyline computation. In: VLDB (2001)
14. Chan, C.-Y., Jagadish, H.V., Tan, K.-L., Tung, A.K.H., Zhang, Z.: On high dimensional skylines. In: Ioannidis, Y., Scholl, M.H., Schmidt, J.W., Matthes, F., Hatzopoulos, M., Böhm, K., Kemper, A., Grust, T., Böhm, C. (eds.) EDBT 2006. LNCS, vol. 3896, pp. 478–495. Springer, Heidelberg (2006)
15. Chan, C.Y., Jagadish, H.V., Tun, K.L., Tung, A.K.H., Zhang, Z.: Finding k-Dominant Skylines in High Dimensional Space. In: SIGMOD (2006)
16. Papadias, D., Tao, Y.: An optimal and progressive algorithm for skyline. In: SIGMOD (2003)
17. Kossmann, K., Ramsak, F., Rost, S.: Shooting Stars in the Sky-An Online Algorithm for Skyline Queries. In: VLDB (2002)

Meta Galaxy: A Flexible and Efficient Cube Model for Data Retrieval in OLAP[*]

Jie Zuo[1], Changjie Tang[1], Lei Duan[1], Yue Wang[1], Liang Tang[1],
Tianqing Zhang[1,**], and Jun Zhu[2]

[1] School of Computer Science, Sichuan University,
Chengdu 610065, China
{zuojie,tangchangjie,zhangtianqing}@cs.scu.edu.cn
[2] National Center for Birth Defects Monitoring
Chengdu 610041, China

Abstract. OLAP is widely used in data analysis. The existing design models, such as star schema and snowflake schema, are not flexible when the data model is changed. For example, the task for inserting a dimension may involve complex operations over model and application implementation. To deal with this problem, a new cube model, called Meta Galaxy, is proposed. The main contributions of this work include: (1) analyzing the shortcoming of traditional design method, (2) proposing a new cube model which is flexible for dimension changes, and (3) designing an index structure and an algorithm to accelerate the cube query. The time complexity of query algorithm is linear. The extensive experiments on the real application and synthetic dataset show that Meta Galaxy is effective and efficient for cube query. Specifically, our method decreases the storage size by 95.12%, decreases the query time by 89.89% in average compared with SQL Server 2005, and has good scalability on data size.

1 Introduction

OLAP provides users quick answers to multi-dimensional analytical queries [1]. The process of OLAP combines technologies of database, data warehouse and data mining. In OLAP, data cubes provide data summaries in different points of view or dimensions. Generally, there are two kinds of OLAP systems: multi-dimensional OLAP (MOLAP) and relational OLAP (ROLAP) [2]. MOLAP and ROLAP are widely used in business related domains. However, both of them have disadvantages respectively. For example, MOLAP approach introduces data redundancy, while ROLAP is not suitable when the model is heavy on calculations which don't translate well into SQL statements.

In the real world applications, the number of dimensions in a data cube may be huge. The traditional design of a data cube creates a dimension table for each dimension.

---

[*] This work was supported by the National Natural Science Foundation of China under grant No. 60773169 and the 11th Five Years Key Programs for Sci. &Tech. Development of China under grant No. 2006BAI05A01.
[**] Correspondence author.

L. Chen et al. (Eds.): APWeb and WAIM 2009, LNCS 5731, pp. 234–244, 2009.
© Springer-Verlag Berlin Heidelberg 2009

However, the dimensions may change due to the new requirements. As a result, a new attribute is added in the facts table, and a new dimension table is created. It follows that the traditional design is not flexible and efficient enough to meet the complex real application requirements. Example 1 illustrates the model design in birth defects monitoring database.

Example 1. In the real application of birth defects monitoring database, each record contains the information of a baby with birth defects as shown in Table 1. All birth defects are list as attributes for a baby record. For each defect attribute, the value describes the status of corresponding defect.

Table 1. Sample records stored in birth defects monitoring database

Baby_id	Weight	Gender	$Defects_1$	$Defects_2$	...	$Defects_n$
001	2800	Male	No	Serious	...	No
002	3500	Female	Weak	Weak	...	No
...	...	...	...	...	...	...

Table 1 shows that a new attribute must be inserted if a new defect is discovered, and needs updating operations over the data model and application design.

To solve the problem, we design a new cube model named Meta Galaxy. Basically, Meta Galaxy consists of three tables:

- Meta-dimension Table. It contains the definitions of all dimensions and uses unique ID to indicate each dimension.
- Meta-item Table. It defines all available values for each dimension.
- Meta-value Facts Table. It records the values on all dimensions of all objects.

The Figure 2, Figure 3 and Figure 4 in the rest sections describe the rough sketch of this model, the skeleton of the model likes a star model (or a galaxy when problem is complex), while the attribute value is meta value, such as *object_id*, *dim_id* and *itm_id*. Hence we call it as Meta Galaxy. It is with following advantages:

- Flexibility. It is unnecessary to modify the model design no matter inserting or deleting a dimension. Additionally, the storage space is saved since many default values are not kept in Meta Galaxy.
- Efficiency. An index structure (called Meta Index) is created and an algorithm is designed to accelerate the data query. Our experiments show that our method decreases the storage size by 95.12%, decreases the query time by 89.89% in average compared with SQL Server 2005, and has good scalability on data size.

2 Related Work

Svetlana et al. applied OLAP to business analysis [3]. The OLAP they implement OLAP is called ROLAP, since it is based on relational DBMS. Based on the concept of emerging pattern, Sébastien et al. introduced the concept of emerging cube, which is materialized in MOLAP, to reveal the changing of trends between two pre-computed cubes [4]. Perez et al. used XML files to store the raw data and established a related cube (R-cube) for MOLAP [5]. Doug et al. extended the OLAP data model to deal with ambiguous, imprecision data, and provided an allocation-based approach to tackle semantic problems in aggregation [6]. In [7], Thomas et al. allocated relational data warehouses based on a star schema and bitmap index, and used a facts table with multi-dimensional hierarchical data fragmentation to support queries referencing subsets of schema dimensions. Zhongzhi Shi et al. investigated the way to map multi-dimensional operations into SQL statements [8]. Zohreh et al. established views and index for OLAP and solved the index-selection problem to improve the processing efficiency [9]. Elaheh et al. designed a novel method, called Partial Pre-aggregation (PP), to estimate approximation results of joint queries in MOLAP [10]. Ruoming Jin et al. optimized the cost of main memory to deal with the performance problems in multi-dimensional hierarchical data [11]. Yon Dohn et al. used the RD-Tree to process top-k queries in OLAP aggregation [12]. T.S. Jung et al. proposed an index structure to accelerate the MOLAP operations [13].

The main differences between our method and existing results are: (a) we pay more attention to the situation of dimension evolutions. (b) It is unnecessary to modify the data model and application implementation due to the proposed Meta Galaxy to handle dimension evolution.

3   The Structure of Meta Galaxy

We propose a new cube model named Meta Galaxy. Intuitively, the design of Meta Galaxy is similar to the Triples. However, Meta Galaxy is a cube model with index and query algorithms (discussed in Section 4). Basically, the Meta Galaxy consists of three tables: Meta-dimension Table, Meta-item Table and Meta-value Facts Table. Figure 1 illustrates the main structures of these three tables.

- The Meta-dimension Table defines all dimensions in the cube model. The attributes include: *dim_id*, *name* and *description*. In this table, attribute *dim_id* is a unique key to indicate each dimension, attribute *name* is the name of a dimension, and attribute *description* is the note for a certain dimension.

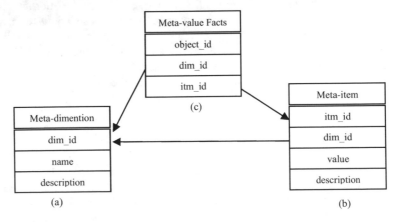

Fig. 1. The relationships among three tables in Meta Galaxy: (a) Meta-dimension Table, (b) Meta-item Table, and (c) Meta-value Facts Table

- The Meta-item Table defines all available values for each dimension. There are four attributes: *itm_id*, *dim_id*, *value*, and *description*. In Meta-item Table, attribute *itm_id* is a unique id for an available value (*value*) of dimension *dim_id*, and attribute *description* is the corresponding note.
- The Meta-value Facts Table (mvFacts) defines values on all dimensions of all objects by recording *object_id*, *dim_id* and *itm_id*.

Example 2. In Example 1, the baby with id "001" is male. Suppose the dimension id (*dim_id*) of "Gender" is *dim003*, item id of (*itm_id*) "Male" is *itm001*. Then, in Meta Galaxy, Meta-dimension Table contains record (*dim003*, "Gender", "*the gender information*"), Meta-item Table contains record (*itm001*, *dim003*, "Male", "*male information*"), and Meta-value Facts Table contains record (*001*, *dim003*, *itm001*).

Since dimension information is recorded in the Meta-dimension Table, it is flexible to handle dimension evolution. For example, to insert a dimension d_1, the record of d_1 will be added in the Meta-dimension table. Furthermore, all available values of d_1 will be added in the Meta-item table. To delete a dimension d_2, all corresponding records related to d_2 in Meta-dimension table and Meta-item table will be removed.

Example 3. Consider adding a new birth defect, *def*, in the birth defect monitoring database. Suppose there are three statuses of *def*: s_1, s_2, and s_3. Then a new tuple containing *def* and related information will be inserted into Meta-dimension table, and three tuples containing *def* and each of its three statuses will be inserted into Meta-item table.

4   The Meta Index and Fast Cube Query Algorithm

4.1   The Design of Meta Index

A typical cube query on a relational data table is a SQL statement containing "group by" clause. For example, Figure 2 describes a typical query statement on dimension d1, d2, d3 with aggregation function agg().

```
SELECT d1, d2, d3, agg()
FROM facts
WHERE dn = value1 and dm = value2
GROUP BY d1, d2, d3
```

Fig. 2. A typical SQL statement for cube query

Figure 2 shows that three key steps are in the query process.

Step 1: Filter objects based on the "where" clause.
Step 2: Group objects based on the "group by" clause.
Step 3: Calculate the results of each group based on the aggregation functions.

The efficiencies of first two steps are important. Since in real OLAP analysis, the number of data size is always very large. In Meta Galaxy, dimensions are not attributes

```
SELECT a, b, c, COUNT(*) FROM
(SELECT m.id AS id, a, b, c FROM
  (SELECT DISTINCT oid AS id FROM mvFacts) m
  LEFT JOIN
  (SELECT object_id AS id, itm_id AS a FROM mvFacts WHERE dim_id = d1) a ON m.id = a.id
  LEFT JOIN
  (SELECT object_id AS id, itm_id AS b FROM mvFacts WHERE dim_id = d2) b ON m.id = b.id
  LEFT JOIN
  (SELECT object_id AS id, itm_id AS c FROM mvFacts WHERE dim_id = d3) c ON m.id = c.id
) AS tt
GOURP BY a, b, c
```

Fig. 3. A modified SQL statement for query in Meta Galaxy

in the Facts Table. Thus to execute queries in Figure 2, we modify the SQL statement for Meta Galaxy. The statements are shown in Figure 3.

Figure 3 illustrates two disadvantages of this SQL statements template:

- It is difficult to write this kind of query statements.
- The efficiency of query is low. The Meta-Value Facts Table should do self-join several times.

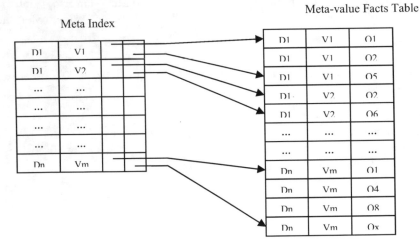

Fig. 4. The index for Meta-value Facts Table

Thus, it is necessary to design query algorithm for Meta Galaxy to improve the efficiency. The key points are: (a) in cube query, all query conditions are presented in the form of "slice". That is, "dimension 1 = value 1 and dimension 2 = value 2 and ... and dimension n = value n". (b) Based on the selection conditions template, we design an efficient index structure which is shown in Figure 4.

The index is a map whose key is a pair (*dim_id*, *itm_id*), and value is a pair (*start*, *end*). The index is constructed in the following way. First, for all records in Meta-value Facts Table, sort them according to (*dim_id*, *itm_id*, *object_id*). As a result, all objects that have the same value on certain dimension will be arranged together. Next, for each key (*dim_id*, *itm_id*) in the index, (*start*, *end*) indicates the first and last positions of records containing (*dim_id*, *itm_id*) in Meta-value Facts Table.

One advantage of adopting this index is saving storage space which can save disk accessing time. The size of index depends on the number of items on all dimensions. Usually, it is small enough to resident in memory. Two key points for Meta-value Facts Table:

(a) All records in the table must be ordered once the index is created.
(b) The attributes *dim_id* and *itm_id* are redundancy. To minimize the storage space and query response time, these two attributes are removed.

As a result, there is only one column in Meta-value Facts Table. That is, the size of Meta-value Facts Table is decided by the number of non-default values of all objects on each dimension.

4.2 The Fast Cube Query Algorithm

We design the cube query algorithm based on the index. The basic idea is filtering objects which do not satisfy the selection conditions and grouping the satisfied objects. Index is used in both filtering and grouping to accelerate the query. We will give the algorithms for filtering and grouping, respectively.

In cube query, typical selection conditions like "dim1 = itm1 and dim2 = itm2...". Based on selection template and Meta Index, we can find the satisfied set of objects by Algorithm 1. The inputs of Algorithm 1 are dimension id and item id given by selection conditions.

Algorithm 1. IndexScan(dim, itm)
Input: dim: the dimension id, itm: the item id
Output: P: the id set of objects whose values on dim are itm

```
begin
1. construct the index key K = <dim, itm>
2. search the index by K, get the value (s, t)
3. scan Meta-value Facts from s to t, get object id set P
4. return P
end.
```

If there are several selection conditions, we use Algorithm 1 to find corresponding objects of each condition, and the final result is their intersection.

The filtering Filter (SC) is based on Algorithm 1.

Algorithm 2. Filter(SC)
Input: SC: the set of selection conditions
Output: P: the id set of objects that satisfy selection conditions

```
begin
1. set P is the id set of all objects
2. for each clause in SC, (dim, itm)
3.     Q ← IndexScan(dim, itm)
4.     P = P ∩ Q
5. return P
end.
```

The Algorithm 2 shows several advantages in the process of finding the objects satis-fying selection conditions: (a) No unrelated object is scanned. (b) The efficiency of disk access is high due to the sequence scan. (c) All IDs got by scan are in sequence. It is efficient to execute intersection operation.

Algorithm 3 gives the details of grouping operation based on the results of filtering step (Algorithm 2).

Algorithm 3. Group(R, D)
Input: R: the id set of objects which satisfy filtering conditions, D: dimension set
Output: P: the id set of objects in each group

```
begin
1.   P ← {R}
2.   for each dimension dim in D
3.      for each available item in dim, itm
4.         T ← IndexScan(dim, itm)
5.         for each id set S in P
6.            Q ← Q ∪ (S ∩ T)
7.         P ← Q
8.      remove all elements in Q
9.   return P
end.
```

Note that, In Step 4, the function IndexScan(dim, itm) returns the id set of objects based on (dim, itm)

The result of Algorithm 3 is the set of groups of object ids which are sorted in lexicographically order. The Algorithm 2 scans Meta-value Facts Table in sequence only once, so the efficiency is high. Since the main part of Algorithm 3, like the Fil-tering Algorithm, is the intersection operation, we apply bit operation to improve the efficiency. Our cube query algorithm as shown in Algorithm 4.

Algorithm 4. CubeQuery(SC, D)
Input: SC: the set of selection conditions, D: the set of dimensions
Output: P: the id set of objects in each group

```
begin
1. R ← Filter(SC)
2. P ← Group(R, D)
3. return P
end.
```

As a cube query may involve the access of data on disk, we should try to minimize the times of disk access. In filtering step (Algorithm 2), our method scans the records related to selection conditions in Meta-value Facts table in sequence only once. In grouping step (Algorithm 3), our method scans the related aggregation dimensions on Meta-value Facts Table in sequence only once. For all meaningful queries, the parts of Meta-value Facts Table scanned in filtering and grouping are different, so the whole Meta-value Facts Table is scanned once in the worst case. As a result, the time complexity of our method is linear. It is worth to note that in most queries of real application, the dimensions involved is a subset of all dimensions, thus only a small part of Meta-value Facts Table will be scanned.

5 Experimental and Performance Study

To evaluate the performance of Meta Galaxy, we test it on the National Birth Defects Monitoring Database of China with about 1,000,000 instances of defect babies. The number of baby information is more than 200, including birth date, gender, body length, weight, parent ages, parent occupations, family history, defect information, and so on. Meta Galaxy is implemented in Java. The experiments are performed on an Intel Core2 1.86 GHz (2 Cores) PC with 4G memory running Windows Server 2008 64 bit Edition operating system.

5.1 Efficiency Comparison

Since the original data are kept in MS SQL Server 2005, we compare our Meta Galaxy with it. We implement two versions of Meta Galaxy, one is implemented based on SQL Server without Meta index, and the other is implemented based on file system with Meta index, denoted as R-Meta Galaxy and F-Meta Galaxy respectively. Table 2 gives the storage size used by each method in our experiment.

Table 2. The storage sizes used by SQL Server, R-Meta Galaxy and F-Meta Galaxy

SQL Server	R-Meta Galaxy	F-Meta Galaxy
2GB	2 GB	100 MB

Table 2 shows that F-Meta Galaxy can save the storage space greatly. Moreover, F-Meta Galaxy is flexible to the change of dimensions. Next, we demonstrate the efficiency of F-Meta Galaxy is higher than other models by four different methods.

- Method 1: Execute SQL statements for cube queries on original data model.
- Method 2: Create a view on R-Meta Galaxy to construct the original data model, and execute queries on this view.
- Method 3: Materialize the view created by Method 2, and execute cube queries after building the index.
- Method 4: Execute cube query algorithm (Algorithm 4) on F-Meta Galaxy to do cube queries.

In Method 2, the view is created by the statements like those in Figure 2.

For each method list above, we execute 4 typical queries. Query 1 is a simple one-dimension grouping query. Query 2 is a group query on three dimensions. And Query 3 and Query 4 are more complex by adding some selection conditions.

- Query 1: SELECT a, count(*) FROM facts GROUP BY a
- Query 2: SELECT a, b, c, count(*) FROM facts GROUP BY a, b, c

- Query 3: SELECT a, b, count(*) FROM facts WHERE c = sc1[1] GROUP BY a, b
- Query 4: SELECT a, b, count(*) FROM facts WHERE c = sc1 AND d = sc2 GROUP BY a, b

Note that, in Method 2, 3 and 4, we slightly revise the statements to adapt to Meta Galaxy, and keep the same semantic meaning. Figure 5 illustrates the query time of each method executing these four queries. The query time of Method 2 is much longer than those of other methods. The query time of Method 2 executing each query is: 6891 ms, 7750 ms, 13947 ms, and 5344 ms, respectively.

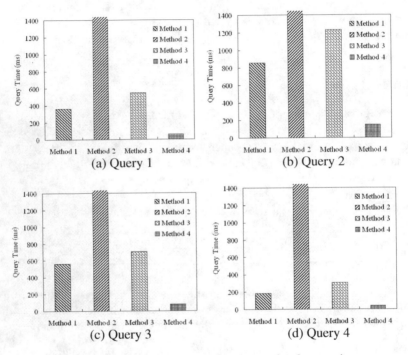

Fig. 5. The query time of each method executing four queries

Figure 5 shows that the proposed method (Method 4) greatly decreases query time. Specifically, compared with Method 1, 2, 3, the query time is decreased by 82.61%, 98.97%, 88.11%, respectively. Moreover, we can see that Method 1 and Method 3 have the similar query performance. The reason is that the query performances on a materialized view and on a table are nearly the same. As our method is implemented in Java, the performance can be improved further if we adopt C++.

5.2 Scalability Validation

To demonstrate the linear time complexity of our method, we record the query time of four queries on different sizes of data. Since the number of records in the National Birth

---

[1] We use sc1 to indicate a selection condition. It is an item id in our experiment.

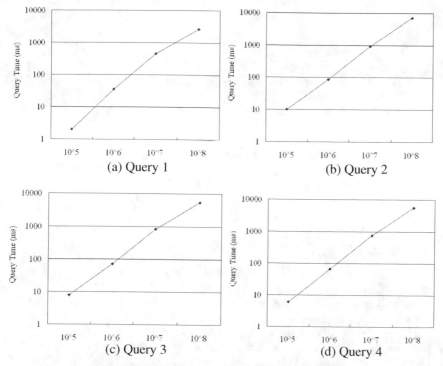

Fig. 6. The query time of four queries in Method 4 under different data sizes

Defects Monitoring Database of China is limited, we built a data generator to randomly generate synthetic datasets. The generated datasets have following characteristics: (a) each dataset contains 200 dimensions, (b) each dimension has 20 items in average, (c) there are 20 dimensions have non-default values in average for each object.

We generate four datasets of which the numbers of objects are: 10^5, 10^6, 10^7, and 10^8. In Meta Galaxy, the storage sizes of these four datasets are: 8M, 80M, 800M and 8G bytes. For each of these datasets, we execute these four queries in Method 4 on it and record the query time. Figure 6 illustrates the query time of each query under different data size. It is clear to see that the query time is increased linearly with the dataset size becomes larger. So our proposed Meta Galaxy has a good scalability.

6 Conclusions

OLAP provide users quick answers to multi-dimensional analytical queries. The traditional design is not flexible when the data model is changed. To deal with this problem, a new cube model, called Meta Galaxy, is proposed. Moreover, an index structure and an algorithm are designed to accelerate the cube query. The time complexity of query algorithm is linear. The extensive experiments demonstrate that the newly proposed method decreases the storage size by 95.12%, decreases the query time by 89.89% in average compared with SQL Server 2005, and has good scalability on data size. The future work includes applying Meta Galaxy to more real world applications.

References

1. Han, J., Kambr, M.: Data Mining Concepts and Techniques. Higher Education Press, Beijing (2001)
2. Pedersen, T.B., Jensen, C.S.: Multidimensional database technology. Computer 34(12), 40–46 (2001)
3. Mansmann, S., Neumuth, T., Scholl, M.H.: OLAP technology for business process intelligence: Challenges and solutions. In: Song, I.-Y., Eder, J., Nguyen, T.M. (eds.) DaWaK 2007. LNCS, vol. 4654, pp. 111–122. Springer, Heidelberg (2007)
4. Nedjar, S., Casali, A., Cicchetti, R., Lakhal, L.: Emerging cubes for trends analysis in OLAP databases. In: Song, I.-Y., Eder, J., Nguyen, T.M. (eds.) DaWaK 2007. LNCS, vol. 4654, pp. 135–144. Springer, Heidelberg (2007)
5. Perez, J.M., Berlanga, R., Aramburu, M.J., Pedersen, T.B.: R-Cubes: OLAP Cubes Contextualized with Documents. In: ICDE (2007)
6. Burdick, D., Deshpande, P.M., Jayram, T.S., Ramakrishnan, R., Vaithyanathan, S.: OLAP over uncertain and imprecise data. The VLDB Journal (2007)
7. Stöhr, T., Märtens, H., Rahm, E.: Multi-Dimensional Database Allocation for Parallel Data Warehouses. In: Proceedings of the 26th VLDB, Cairo, Egypt (2000)
8. Shi, Z., Huang, Y., He, Q., Xu, L., Liu, S., Qin, L., Jia, Z., Li, J., Huang, H., Zhao, L.: MSMiner—a developing platform for OLAP. Decision Support Systems, 42 (2007)
9. Asgharzadeh, Z., Chirkova, R., Fathi, Y.: Exact and Inexact Methods for Selecting Views and Indexes for OLAP Performance Improvement. In: EDBT (2008)
10. Pourabbas, E., Shoshani, A.: Efficient Estimation of Joint Queries from Multiple OLAP Databases. ACM Trans. on Database Systems V(N) (September 2006)
11. Jin, R., Yang, G., Agrawal, G.: Communication and Memory Optimal Parallel Data Cube Construction. IEEE Trans. on Parallel and Distributed Systems 12 (2005)
12. Chung, Y.D., Yang, W.S., Kim, M.H.: An efficient, robust method for processing of partial top-k/bottom-k queries using the RD-Tree in OLAP. Decision Support Systems 43 (2007)
13. Jung, T.S., Ahn, M.S., Cho, W.S.: An Efficient OLAP Query Processing Technique Using Measure Attribute Indexes. In: WISE 2004, pp. 218–228 (2004)

Author Index